NEW YORK

Rocks & Minerals

A Field Guide to the Empire State

Dan R. Lynch & Bob Lynch

Adventure Publications
Cambridge, MN

Dedication

To Nancy Lynch, wife of Bob and mother of Dan, for her love and continued support of our book projects.

And to Julie Kirsch, Dan's lovely wife, for her seemingly endless patience and enthusiasm for all these books we keep making.

Acknowledgments

Thanks to the following for providing specimens and/or information: Rob Rosenblatt (Rocko Minerals), George Robinson, Ph.D., Susan Robinson, Ted Smith, and Kean Riley

Photography by Dan R. Lynch

Cover and book design by Jonathan Norberg

Edited by Brett Ortler

10 9 8 7 6 5 4 3 2 1

Published by Adventure Publications
820 Cleveland Street South
Cambridge, MN 55008
1-800-678-7006
www.adventurepublications.net

Printed in China
ISBN: 978-1-59193-524-7

Table of Contents

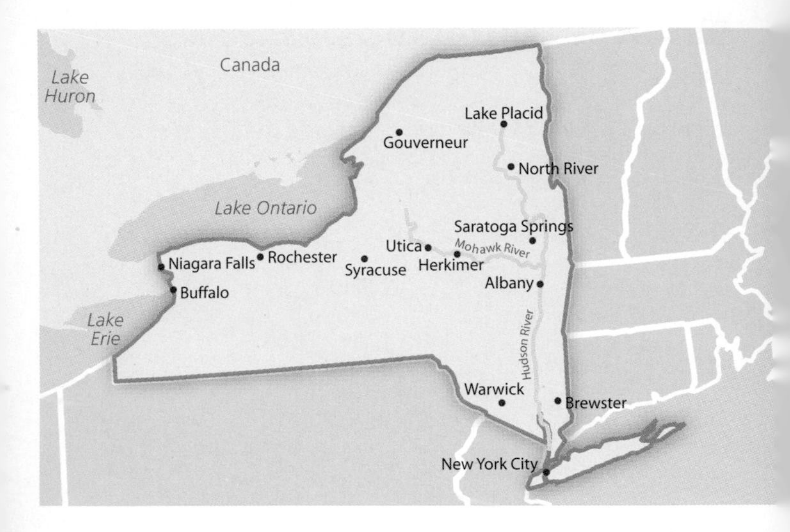

Introduction

The northeastern portion of the United States is home to a diverse array of geological sites and an incredible history, and New York State holds a prominent place in both. Collectors from around the country (and the world) have long looked to New York for rare mineral specimens, scientifically important fossils, and valuable ores. While most people imagine New York City when they think of New York, mineral collectors often leave the urban sprawl behind and head upstate where the mountains and rivers reveal the treasures just beneath your feet. This book will discuss many of the rocks and minerals that this geologically stunning state has to offer, including everything from its most common rocks to its rarest crystals. More importantly, we'll tell you what to look for and how to identify what you've found.

Important Terms and Definitions

Books that cover topics in geology are typically not written so that amateur rock-collecting hobbyists can understand them. So in order to make this book intuitive for novices yet still useful for experienced collectors, we have included a few technical geological terms in the text, but we "translate" them immediately after using them by providing a brief definition. In this way, amateurs can learn some of the more important terms relevant to the hobby in an easy, straightforward manner. Of course, all of the geology-related terms used here are defined in the glossary found at the back of this book as well. But for those entirely new to rock and mineral collecting, there are a few very important terms you should understand not only before you begin researching and collecting minerals, but even before you read this book.

Many people go hunting for rocks and minerals without knowing the difference between the two. But the difference is simple: a **mineral** is formed from the crystallization (solidification) of a chemical compound, or combination of elements. For example, silicon dioxide, a chemical compound consisting of the elements silicon and oxygen, crystallizes to form quartz, the most abundant mineral on earth. In contrast, a **rock**, is a mass of solid material containing a mixture of many different minerals. While pure minerals exhibit very definite and testable characteristics, such as a distinct repeatable shape and hardness, rocks do not and vary greatly because of the various minerals contained within them. This can make identifying rocks more difficult for amateurs than identification of minerals.

Many of the important terms critical to rockhounds apply only to minerals and their crystals. A **crystal** is a solid object with a distinct shape and a repeating atomic structure created by the solidification of a chemical compound. In other words, when different elements come together, they form a chemical compound that will take on a particular shape when it hardens. For example, the mineral galena is lead sulfide, a chemical compound consisting of lead and sulfur, which **crystallizes**, or solidifies, into the shape of a cube. A "repeating atomic structure" means that when a crystal grows, it builds upon itself. If you compared two crystals of galena, one an inch long and the other a foot long, they would have the same identical cubic shape. In contrast, if a mineral is not found in a well-crystallized form but rather as a solid, rough chunk, it is said to be **massive**. If a mineral typically forms **massively**, it will frequently be found as irregular pieces or masses, rather than as well-formed crystals.

Cleavage is the property of some minerals to break in a particular way when carefully struck. As solid as minerals may seem, many have planes of weakness within them that derive from a mineral's internal crystal structure. These points of weakness are called **cleavage planes** and it is along these planes that some minerals will **cleave**, or separate, when struck. For example, galena has cubic cleavage, and even the most irregular piece of galena will fragment into perfect cubes if carefully broken.

Luster is the intensity with which a mineral reflects light. The luster of a mineral is described by comparing its reflectivity to that of a known material. A mineral with "glassy" luster (also called "vitreous" luster), for example, is similar to the "shininess" of glass. The distinction of a "dull" luster is reserved for the most poorly reflective minerals, while "adamantine" describes the most brilliant. Minerals with a "metallic" luster clearly resemble metal, and this can be a very diagnostic trait. But determining a mineral's luster is a subjective experience, so not all observers agree, especially when it comes to less obvious lusters such as "waxy," "greasy," and "earthy."When minerals form, they do so on or in rocks. Therefore, it is important to understand the distinction between the different types of rocks if you hope to successfully find a specific mineral. **Igneous** rocks form as a result of volcanic activity and originate from magma, lava, or volcanic ash. **Magma** is hot, molten rock buried deep within the earth and can take extremely long periods of time to cool and form rock. **Lava**, on the other hand, is molten rock that has reached the earth's surface where it cools and solidifies into rock very rapidly. **Sedimentary** rocks typically form at the bottoms of lakes and oceans when sediment compacts and solidifies into layered masses. This sediment can contain organic matter as well as weathered fragments or grains from broken-down igneous rocks, metamorphic rocks, or other sedimentary rocks. Finally, **metamorphic** rocks develop when igneous, sedimentary, or even other metamorphic rocks are subjected to heat and/or pressure within the earth and are changed both in appearance and mineral composition.

A Brief Overview of New York's Geology

To the untrained eye, the Northeast's geology is rather perplexing; seafloors are found on mountaintops, very old rocks tower above much younger rocks, and visitors can find mountain ranges of markedly different ages and makeups standing nearly shoulder-to-shoulder. Indeed, it is the Northeast's mountains, perhaps more than anything else, that tell us much of how the region came to be, and New York State is at the heart of the story.

When discussing the geological history of a region, we define it by its bedrock, which is the uppermost layer of solid rock just beneath the loose soil and gravel on the surface. Bedrock forms in layers, often composed of completely different types of rocks, and the youngest layers form above the older layers buried deeply below. Subsequent weathering erodes the upper, younger layers and reveals the older layers, thus giving us a view of the deeper layers while also revealing clues about the past events that caused the erosion. By looking at the composition of bedrock layers, the amount and patterns of erosion, and the placement of different kinds of bedrock throughout a region, geologists can begin to piece together the origin of landforms and decipher the history of geological events that took place there.

The Adirondack Mountains, typically considered part of the Appalachian Mountains, are the logical starting point in New York's history, as they contain the oldest rocks in New York, and in the entire Northeastern United States. The Adirondacks formed around one billion years ago when all of the earth's continents were in a single mass known as Rodinia. Their constituent rocks were primarily igneous in origin, in the form of both intrusive volcanic rocks, primarily anorthosite, and extrusive rocks. (Extrusive rocks formed on the surface of the earth, whereas intrusive igneous rocks only form deep within the earth's crust.) Most of this material remained deep within the earth for almost its entire lifetime, where it was subjected to increasing pressure and heat as millions of years of sediment weighed down upon it, causing it to undergo considerable metamorphism. When two neighboring tectonic plates (the enormous sheets of mobile rock that make up the earth's crust) collided just 20 million years ago, these rocks were finally forced upwards, leading to the rapid erosion of the overlying sedimentary rocks. This process revealed the present-day arrangement of an anorthosite core sheathed in gneiss and other metamorphic rocks, bounded on all sides by the much younger sedimentary rocks that dominate much of the state. Interestingly, the Adirondacks are actually still rising at an estimated 1 to 3 millimeters per year, making them a relatively young mountain range but one that consists of remarkably old rocks.

As the ancient supercontinent of Rodinia broke up, more volcanic rocks formed atop the then-buried Adirondack rocks, and around 550 million years ago New York was situated on the coast of an ancient ocean. Less than 10 million years later, late in the geological period called the Cambrian, rising sea levels submerged New York beneath tropical seas that teemed with life. Over time, enormous amounts of limestone formed, preserving fossil evidence of coral reefs and other organisms. Then about 100 million years later, during the Ordovician period, a tectonic plate collision, much like the one that would cause the Adirondacks to rise from the earth more than 400 million years later, caused sandstone and other sedimentary rocks on the seafloor to uplift, forming a long, narrow ridge running more-or-less parallel to the modern shoreline of the East Coast. Today, this ridge is known as the Taconic Mountains of eastern New York. These ancient mountains, their peaks rounded and smoothed by erosion, consist largely of metamorphic rocks that formed during the plate collision, which produced immense heat and pressure and altered the extant sedimentary rocks. But the formation of the Taconics disturbed the state's warm seas, ending the deposition of limestone and replacing it with that of shale, mud, and sand. Then, at the end of the Ordovician, the seas retreated entirely, leaving shallow sandstone formations over much of the state.

As with many other states, the inland seas that flooded the landscape weren't gone for long. Around 430 million years ago, sea levels rose once again, submerging the state and leading to the formation of shales, limestones, and other fossil-rich sedimentary rocks. This sea was highly saline, which is why we find halite (also known as common salt) throughout New York State, but that didn't prevent enormous coral reefs from forming. Today, one of the most spectacular reminders of this period is Niagara Falls, which tumble over cliffs of coral reef-bearing dolostone, which is a variety of limestone.

The next 20 million years saw the recession of one sea and the advance of yet another during the Devonian period. The new body of water was an enormous tropical sea extending into the Midwest, and it deposited limestone in large amounts for the last time in New York's history. This was an active period for the region, geologically speaking, as volcanoes in nearby Maine spewed lava and ash, and rivers from the east dumped enormous amounts of sand, mud, and other sediments on the coasts of the sea. As these formations of sediment grew to be thousands of feet thick, the increasing pressure began to solidify them into bodies of rock. Thus began the first step in the formation of the Catskill Mountains.

The next two geological periods were the Carboniferous and Permian, from around 362 to 245 million years ago, but rocks from this time are nearly completely missing from the state, not because they didn't form, but because they've all but eroded away. The modern-day Catskill Mountains provide evidence that these rocks were indeed once present, as we can tell from the mountains' current carbon deposits that they were once buried under approximately four miles of rock. So we know an incredible amount of erosion took place, and in fact, this the reason why the Catskill Mountains exist at all. The sedimentary rocks of the Catskills were never uplifted like the rocks of the nearby mountain ranges; instead, they existed as a large, flat plateau that was carved into distinct peaks and valleys by rivers and faults. This type of landform is described as a dissected plateau, meaning that they aren't technically mountains, geologically speaking. The Catskills are the easternmost portion of a much larger plateau structure called the Allegheny Plateau, which dominates much of southern and western New York's geography.

As dinosaurs crossed the region during the Triassic and Jurassic, 245 to 144 million years ago, they left their footprints, still visible in New York today, in the sediments of the arid environments present at the time. Around the same time, the Atlantic Ocean was beginning to form on the nearby coasts. During the Cretaceous, 144 to 65 million years ago, modern-day Staten Island and Long Island were already on the Atlantic coast, and much of the region's geology was in place.

The last dramatic event in New York's geologic history were the ice ages that affected North America. The most recent ice age, lasting from 110,000 years ago through today, saw glaciers descend from Canada and cover many northern states under immense amounts of ice, including almost the entirety of New York, until they began to recede just 10,000 years ago. Glacial ice moves and flows, albeit very slowly, pushing, scouring, and scraping the rocks it slides upon, essentially "fast-forwarding" erosion, grinding away miles of rock, smoothing mountains, and gouging out valleys. Glaciers can strip enormous amounts of rock from the landscape, grinding it into gravel and sand that adds to their abrasive effect before being dumping the material into rivers and lakes. In many regions, it is the glaciers that are responsible for the formation of many rivers, but in New York, most major rivers existed before the glacial period and were merely altered by the ice. The Finger Lakes in central New York, for example, were extant river valleys that were made deeper by the glaciers; similarly, the Mohawk and Hudson Rivers were already present, but made deeper and more varied by the scouring ice. The recession of the glaciers left New York State with the landforms, rivers, and lakes we see today, but underneath the verdant forests and bustling cities lie the ancient reefs, riverbeds, and unimaginably old igneous rocks that give New York its incredible geological character.

Precautions and Preparations

Rock and mineral collecting often brings with it several dangers and legal concerns. It is always your responsibility to know where you can legally collect, which minerals may be hazardous to your health, and what you need to take with you in order to be prepared for any difficulties you may face.

Protected and Private Land

New York has nationally protected parks and monuments as well as American Indian reservations, all of which are areas where it is illegal to collect anything. In the region's many state parks, collecting is prohibited. We encourage collectors to obey the law and leave the designated natural spaces wild and untouched for generations to come. It is always your responsibility to know whether or not the area in which you are collecting is protected.

As in any state, many places in New York are privately owned, including areas of wilderness that may not have obvious signage. Needless to say, you are trespassing if you collect on private property and the penalty may be worse than just a fine. In addition, property lines change frequently, as do their owners, so just because a landowner gave you permission to collect on their property last year doesn't mean the new owner will like you on their property this year. Always be aware of where you are.

Dangers of Quarries, Rivers, and Sinkholes

When in the field, vigilance and caution are key to remaining safe. Many quarries, gravel pits, and mine dumps present amazing collecting opportunities, but they may have large rock piles or pit walls that are unstable and prone to collapse. Never go beneath overhanging rock, and keep clear of unstable rock walls. Rivers also present their own dangers, even though the water's surface may look calm, strong currents may be present. It doesn't take very much water current to make you lose your footing completely. Finally, in limestone-rich areas such as New York, groundwater dissolving areas of surface rock may develop sinkholes and other pitted features that may be hidden by grass and other plants and present a tripping or falling hazard. In all cases, vigilance and care will ensure a safe collecting trip.

Equipment and Supplies

When you start collecting rocks and minerals, there are a few items you shouldn't forget. No matter where you are collecting, leather gloves are a good idea, as are kneepads if you plan to spend a lot of time on the ground. If you think you'll be breaking rock, bring your rock hammer (not a nail hammer) and eye protection. If the weather is hot and sunny, take the proper precautions and use sunblock and wear sunglasses and a hat. Also, bring ample water, both for drinking and for rinsing specimens. Lastly, bringing a global positioning system (GPS) device is a great way to prevent getting lost.

Collecting Etiquette

Too often, popular collecting sites are closed by landowners due to litter, trespassing, and vandalism. In many of these cases, landowners may have been kind enough to allow collectors onto their land, but when people trespass rather than simply ask for permission, then we all lose. When collecting, never go onto private property unless you've obtained permission, and always be courteous; if you don't dig indiscriminately and don't take more than you need, you're likely to be invited back. To ensure great collecting sites for future rock hounds, dig carefully, take only a few specimens, and leave the location cleaner than you found it.

Dangerous Minerals and Protected Fossils of New York State

Potentially Hazardous Minerals

The vast majority of minerals in New York are completely safe to handle, collect, and store, but a few do have some inherent dangers associated with them. Potentially hazardous minerals included in this book are identified with the symbol shown above. Always take proper precautions with these minerals.

Amphibole group (page 39)—a few varieties are asbestos; asbestiform minerals form as delicate, flexible fiber-like crystals that can easily become airborne and inhaled, posing a cancer risk; these varieties are uncommon

Arsenopyrite (page 51)—contains arsenic, a toxin; wash hands after handling

Cerussite (page 69)—contains lead, a toxin; wash hands after handling

Galena (page 113)—contains lead, a toxin; wash hands after handling

Löllingite (page 153)—contains arsenic, a toxin; wash hands after handling

"Mountain Leather" (page 167)—typically considered a variety of asbestos. Avoid inhaling dust; wear a respirator.

Serpentine group (page 199)—a few varieties are asbestos, and can be dangerous. Wear a respirator and avoid all dust created by serpentine minerals.

It should be noted that collectors don't need to shy away from these minerals; you just need to be mindful of their potential hazards. Galena, for example, may contain lead, but the lead atoms are bonded to sulfur atoms, making the mineral safe to handle—but you probably will still want to wash your hands afterwards. With these minerals (and most other minerals in general), the real danger is in inhaling their dust or ingesting them, which is easily avoided.

Protected Fossils and Artifacts

Fossils are the remains of ancient plants and animals that have turned to rock, and they are extremely popular collectibles. In most states, including New York, there are strict rules about fossil collecting. Vertebrate fossils (fossils of animals with a backbone) are protected due to their possible scientific significance. These should be reported to the Bureau of Land Management, and if you collect such fossils you may incur fines or other penalties. In general, on state-owned land, permits must be acquired for any kind of collecting. For more information, contact the Bureau of Land Management's Northeastern States Field Office by calling 414-297-4400 or visiting www.blm.gov. In addition, you may find artifacts created by American Indians in New York, including arrowheads, pottery, and pottery fragments, but it is illegal in all cases to disturb or collect these artifacts! They may hold considerable scientific and cultural value. If you come across an artifact, contact the Bureau of Land Management.

Hardness and Streak

There are two important techniques everyone wishing to identify minerals should know: hardness and streak tests. All minerals will yield results in both tests, as will certain rocks, which makes these tests indispensable to collectors.

The measure of how resistant a mineral is to abrasion is called hardness. The most common hardness scale, called the Mohs Hardness Scale, ranges from 1 to 10, with 10 being the hardest. An example of a mineral with a hardness of 1 is talc; it is a chalky mineral that can easily be scratched by your fingernail. An example of a mineral with a hardness of 10 is diamond, which is the hardest naturally occurring substance on earth and will scratch every other mineral. Most minerals fall somewhere in the range of 2 to 7 on the Mohs Hardness Scale, so learning how to perform a hardness test (also known as a scratch test) is critical. Common tools used in a hardness test include your fingernail, a U.S. nickel (coin), a piece of glass and a steel pocket knife. There are also hardness kits you can purchase that have a tool of each hardness.

To perform a scratch test, you simply scratch a mineral with a tool of a known hardness—for example, we know a steel knife has a hardness of about 5.5. If the mineral is not scratched, you will then move to a tool of greater hardness until the mineral is scratched. If a tool that is 6.5 in hardness scratches your specimen, but a 5.5 did not, you can conclude that your mineral is a 6 in hardness. Two tips to consider: As you will be putting a scratch on the specimen, perform the test on the backside of the piece (or, better yet, on a lower-quality specimen of the same mineral), and start with tools softer in hardness and work your way up. On page 18 you'll find a chart that shows which tools will scratch a mineral of a particular hardness.

The second test every amateur geologist and rock collector should know is streak. When a mineral is crushed or powdered, it will have a distinct color—this color is the same as the streak color. When a mineral is rubbed along a streak plate, it will leave behind a powdery stripe of color, called the streak. This is an important test to perform because sometimes the streak color will differ greatly from the mineral itself. Hematite, for example, is a dark, metallic and gray mineral, yet its streak is a rusty red color. Streak plates are sold in some rock and mineral shops, but if you cannot find one, a simple unglazed piece of porcelain from a hardware store will work. But there are two things you need to remember about streak tests: If the mineral is harder than the streak plate, it will not produce a streak and will instead scratch the plate itself. Secondly, don't bother testing rocks for streak; they are made up of many different minerals and won't produce a consistent color.

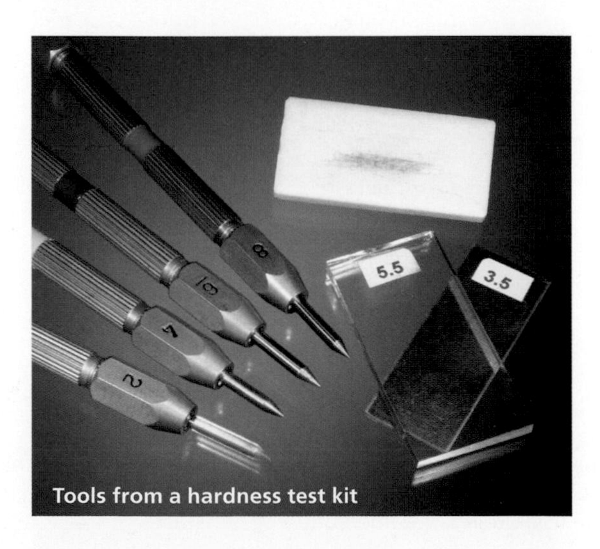

Tools from a hardness test kit

The Mohs Hardness Scale

The Mohs Hardness Scale is the primary measure of mineral hardness. This scale ranges from 1 to 10, from softest to hardest. Ten minerals commonly associated with the scale are listed here as well as some common tools used to determine a mineral's hardness. If a mineral is scratched by a tool of a known hardness, then you know it is softer than that tool.

HARDNESS	EXAMPLE MINERAL	TOOL
1	Talc	
2	Gypsum	
2.5		Fingernail
3	Calcite	
3.5		U.S. nickel, brass
4	Fluorite	
5	Apatite	
5.5		Glass, steel knife
6	Orthoclase feldspar	
6.5		Streak plate
7	Quartz	
7.5		Hardened steel file
8	Topaz	
9	Corundum	
9.5		Silicon carbide
10	Diamond	

For example, if a mineral is scratched by a U.S. nickel (coin) but not your fingernail, you can conclude that its hardness is 3, equal to that of calcite. If a mineral is harder than 6.5, or the hardness of a streak plate, it will instead scratch the streak plate itself, unless weathered to a softer state.

Quick Identification Guide

Use this quick identification guide to help you determine which rock or mineral you may have found. Listed here are the primary color groups followed by some basic characteristics of the rocks and minerals of New York as well as the page number where you can read more. While the most common traits for each rock or mineral are listed here, be aware that your specimen may differ greatly.

WHITE OR COLORLESS

If white or colorless and...	then try...
Crusts or radial groupings of needle-like crystals in limestone	aragonite, page 49
"Fuzzy" groupings of tiny needle-like crystals on serpentine	artinite, page 53
Blocky or plate-like crystals that are very heavy for their size	baryte, page 55
Common, soft, six-sided tooth-like crystals or blocky masses embedded in limestone	calcite, page 63
Blocky, angular crystals grown in parallel groups, sometimes pale blue	celestine, page 67
Soft, often powdery crusts on galena, sometimes gray or tan	cerussite, page 69
Rare, brightly lustrous, glassy angular crystals in basalt or diabase	datolite, page 83

Quick Identification Guide (continued)

(continued) **If white or colorless and...**	**then try...**
Soft, blocky, rhombohedral crystals, often with curved faces and a pearly luster in limestone	dolomite, page 87
Hard, glassy, reflective, blocky masses found in rocks, particularly granite	feldspar group, page 93
Glassy, often very transparent cubic crystals in limestone	fluorite, page 95
Glassy or fibrous and extremely soft crystals or masses in sedimentary rocks	gypsum, page 125
Glassy, translucent, blocky masses that quickly dissolve in water	halite, page 127
"Fuzzy" or chalky crusts on serpentine, often with artinite	hydromagnesite, page 139
Coarsely grained, lustrous, soft rock with the properties of calcite	marble, page 157
Hard, blocky masses that exhibit a white internal schiller under bright light	"moonstone," page 219
Hard, abundant, six-sided pointed crystals or light-colored masses embedded in rocks	quartz, page 185

Quick Identification Guide (continued)

WHITE OR COLORLESS

(continued)	If white or colorless and...	then try...
	Fairly hard blocky opaque crystals embedded in metamorphic rocks, often with pyroxene group minerals	scapolite group, page 193
	Rare clusters of tiny needle-like crystals that are typically fluorescent; found in sedimentary rocks	strontianite, page 217
	Extremely soft, "soapy" feeling masses or flaky crystals easily scratched with a fingernail	talc/soapstone, page 221
	Fibrous, silky masses containing elongated crystals tightly intergrown together	tremolite, page 229
	Soft, silky masses containing parallel fiber-like crystals; often fluorescent	wollastonite, page 243
	Tiny, soft, glassy or pearly crystals of various shapes within cavities in basalt or diabase	zeolite group, page 245

GRAY

	If gray and...	then try...
	Coarse-grained rock with many light-colored, glassy, translucent masses and crystals	anorthosite, page 43
	Uncommon, dark, fine-grained dense rocks, sometimes with vesicles (gas bubbles)	basalt/diabase, page 57

Quick Identification Guide (continued)

(continued) If gray and...	then try...
Very hard, translucent masses, often with a waxy luster; typically irregular in shape	chalcedony, page 71
Very hard, abundant, opaque masses of rock with waxy surfaces when weathered; breaks with sharp edges	chert, page 77
Hard, blocky masses or square crystals embedded in metamorphic rocks, particularly marble	diopside, page 85
Very rare fossil creature resembling a crab or scorpion embedded in shale	fossils, eurypterids, page 103
Shiny, soft, nearly metallic, flexible crystals or masses embedded in coarse-grained rocks or schist	mica group, page 161
Very hard, glassy rock that resembles quartz but has a grainy texture and a flaky appearance when broken	quartzite, page 189
Soft, fine-grained rock consisting of flat, parallel layers that can typically be separated with a knife	shale, page 201
Tiny, dark, dull, elongated crystals embedded in marble from the Warwick area	warwickite, page 239

Quick Identification Guide (continued)

BLACK

If black and...	then try...
Hard grains or masses, often lustrous with a silky sheen; found embedded in coarse-grained rocks	amphibole group, page 39
Very hard, opaque masses of rock with waxy surfaces when weathered; produces a spark when struck with steel	chert, page 77
Soft, sometimes sticky crusts or coatings of oil-like material in limestone or on minerals	hydrocarbons, page 137
Dark, lustrous feldspar mineral that exhibits a blue-green/orange schiller when rotated in bright light	labradorite, page 145
Hard, glassy, dark masses or crystals embedded in rocks, especially coarse rocks like gabbro and pegmatite	pyroxene group, page 181
Hard, octahedral (eight-faced) crystals embedded in metamorphic rocks or quartz	spinel group, page 215
Brightly lustrous, hard, blocky crystals embedded in calcite in metamorphic rocks	tourmaline group, page 227

Quick Identification Guide (continued)

If tan to brown and...	then try...
Uniquely round ball- or blob-like rocky masses found in or near sedimentary rocks	concretions, page 81
Sedimentary rocks, primarily shale or limestone, that contain traces of ancient life such as shells	fossils, page 97
Sedimentary rocks containing clam-like shells or circular, tubular structures	fossils, aquatic animals, page 99
Sedimentary rocks containing gauze-like textures resembling coral	fossils, colonial animals, page 101
Shale or sandstone containing unusual indentations, channels, or tunnels	fossils, trace, page 107
Sedimentary rocks containing fragments of washboard-like segmented shells	fossils, trilobites, page 109
Very hard, opaque masses with a rough texture when broken but a smooth, waxy look and feel when weathered	jasper, page 143
Soft, abundant rock that can be easily scratched with a knife and fizzes slightly in vinegar	limestone/dolostone, page 149
Shiny, soft, nearly metallic, flexible crystals or masses embedded in coarse-grained rocks or schist	mica group, page 161

Quick Identification Guide (continued)

TAN TO BROWN

(continued) If tan to brown and...	then try...
Flexible mats of fibrous mineral material that resemble fabric	"mountain leather," page 167
Extremely fine-grained rocks that are dense, gritty, and somewhat soft; similar to shale	mudstone/siltstone, page 169
Rough, sometimes crumbly, rock composed of tiny grains of sand cemented together	sandstone, page 191
Soft, sometimes pearly, rhombohedral or bladed crystals within cavities in iron-rich rocks	siderite, page 203
Hard, fibrous radiating or parallel clusters of silky crystals in metamorphic rocks such as schist	sillimanite, page 205
Very rare, very hard, inconspicuous grains embedded in marble or skarn, often with serendibite	sinhalite, page 207
Glassy, lustrous wedge-shaped crystals embedded in coarse-grained rocks such as pegmatite	titanite, page 225
Rare, glassy, translucent crystals with a square cross section found embedded in metamorphic rocks	vesuvianite, page 237

Quick Identification Guide (continued)

If green to blue and...	then try...
Hard, striated (grooved) crystals, sometimes translucent, found embedded in metamorphic rocks	actinolite, page 37
Moderately hard, glassy, translucent, six-sided crystals found in very coarse-grained rocks such as pegmatite	apatite group, page 47
Short, blocky, glassy translucent crystals, often in parallel groups, embedded in limestone	celestine, page 67
Hard, blocky crystals with a square cross section; abundant in metamorphic rocks	diopside, page 85
Hard, yellow-green, grooved crystals, crusts, or masses, typically glassy and translucent	epidote, page 91
Very hard, opaque masses with a rough texture when broken but a smooth, waxy look and feel when weathered	jasper, page 143
Very rare, soft, mostly blue rock with white and brassy spots	lapis lazuli/lazurite, page 147
Tiny needle-like crystals arranged in radiating groups, within cavities in iron-rich rock	pecoraite, page 173
Hard, small masses embedded in marble or skarn	serendibite, page 197

Quick Identification Guide (continued)

GREEN TO BLUE

(continued) If green to blue and...	then try...
Soft masses of dark material that feel "greasy" to the touch, sometimes layered or fibrous	serpentine group, page 199
Fibrous, silky masses containing elongated crystals tightly intergrown together, often parallel	tremolite, page 229
Rare, elongated rectangular crystals grown in clusters, often pitted and rough	uralite, page 235

YELLOW

If yellow and...	then try...
Hard, elongated, glassy six-sided crystals with flattened tips, found embedded in very coarse rocks	beryl, page 59
Very rare, extremely hard glassy masses embedded in quartz within pegmatites	chrysoberyl, page 79
Very rare, tiny, flat, striated (grooved) crystals embedded in rock from the Balmat-Edwards area	donpeacorite, page 89
Common coatings, crusts, or veins of soft, rusty material with no crystal shape	limonite, page 151
Hard, glassy, translucent grains or masses; typically in coarse rocks like gabbro or loose in sand	olivine group, page 171

Quick Identification Guide *(continued)*

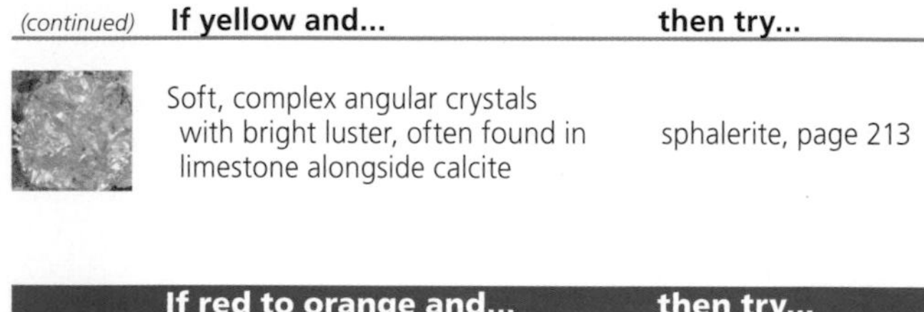

(continued) **If yellow and...**	**then try...**
Soft, complex angular crystals with bright luster, often found in limestone alongside calcite	sphalerite, page 213

If red to orange and...	**then try...**
Very hard, translucent masses with a waxy luster and feel, often found in rivers or on beaches	chalcedony, page 71
Blocky, angular, opaque crystals or masses embedded in coarse-grained rocks like granite	feldspar group, page 93
Very hard, glassy crystals resembling faceted balls, often embedded in schist or coarse-grained rocks	garnet group, page 115
Common crusts or masses, sometimes with a metallic luster or with a chalky texture	hematite, page 129
Glassy, translucent, often isolated grains or masses embedded in metamorphic rocks	humite group, page 135
Soft, complexly shaped crystals with bright luster, often in limestone with dolomite	sphalerite, page 213
Granular feldspar masses containing colorful flecks or flakes within	"sunstone," page 219

Quick Identification Guide (continued)

VIOLET TO PINK

If violet to pink and...	then try...
Soft, glassy, cubic or eight-faced crystals, often in limestone with calcite or hydrocarbons	fluorite, page 95
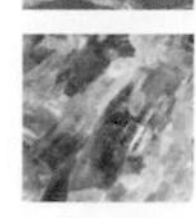 Very hard, glassy, translucent masses found in pegmatite	quartz, varieties, page 187
Fibrous, silky masses containing elongated crystals tightly intergrown together, sometimes parallel	tremolite, varieties, page 231

METALLIC

If metallic and...	then try...
Fairly hard, silvery masses or elongated crystals with a diamond-shaped cross section embedded in marble	arsenopyrite, page 51
Tiny flake- or plate-like crystals arranged into radiating clusters in iron-rich rock	chalcodite, page 73
Fairly soft, brassy masses or triangular crystals, often with sphalerite	chalcopyrite, page 75
Very heavy metallic gray mineral occurring as cubic or blocky masses; often coated with cerussite	galena, page 113
Very soft, flexible, six-sided, thin, flaky crystals, often embedded in marble	graphite, page 121

Quick Identification Guide *(continued)*

(continued) If metallic and...	then try...
Very rare, tiny black metallic needles in talc from St. Lawrence County	groutite, page 123
Common gray metallic material forming bladed crystals, masses or crusts, often with reddish surfaces	hematite, page 129
Black, lustrous masses or grains that are slightly magnetic	ilmenite, page 141
Very rare, brittle, bright silvery mineral grains or elongated crystals in metamorphic rocks	löllingite, page 153
Metallic black masses or eight-faced crystals that bond strongly to a magnet	magnetite, page 155
Hard, brittle, brassy crystal groups with flattened, blade-like or needle-like crystals; often on dolomite	marcasite, page 159
Thin, hair-like brassy needles in radiating clusters in iron-rich rock	millerite, page 165
Hard, brassy yellow to brown metallic cubes or ball-like crystals; pyrite is the most common brassy mineral	pyrite, page 177
Brassy brown six-sided crystals or masses that are magnetic	pyrrhotite, page 183

Quick Identification Guide (continued)

METALLIC

(continued) If metallic and...	then try...
Rare, small, black crystals shaped like triangular pyramids	tetrahedrite group, page 223

MULTICOLORED; LAYERED OR MOTTLED

If multicolored; layered or mottled and...	then try...
Grainy rock with green and tan layers, often quite dark in color; found in the Adirondack Mountains	amphibolite, page 41
Coarse-grained rock containing large bluish feldspar crystals and tiny red garnets; found in the Adirondacks	anorthosite, metamorphosed, page 45
Rock consisting of smaller angular or rounded rocks cemented together	breccia/conglomerate, page 61
Soft sedimentary rock containing countless small aquatic animal fossils such as shells	fossils, reef, page 105
Dark, dense, coarse-grained rock containing large glassy crystals and grains; found in the Adirondacks	gabbro, page 111
Generally hard, dense rock containing coarse, poorly defined layers, often with embedded red garnets	gneiss, page 117
Abundant, coarse-grained rock with mottled coloration; generally light-colored with small dark spots	granite, page 119

Quick Identification Guide (continued)

— MULTICOLORED; LAYERED OR MOTTLED —

(continued)	If multicolored; layered or mottled and...	then try...
	Extremely coarse-grained rock containing many large masses or crystals of many different minerals	pegmatite, page 175
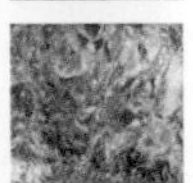	Fairly hard, dense rocks containing many tightly compressed layers; often rich with mica minerals	schist, page 195
	Coarse-grained rock containing many well-developed crystals; often tightly compact with finer grained areas	skarn, page 209
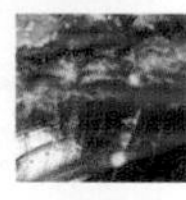	Glass-like masses with varying color and banding, often containing air bubbles within; man-made glass	slag, page 211

— FLUORESCENT —

	If primarily fluorescent and...	then try...
	Tan rock from the Balmat-Edwards area mines that fluoresces vivid orange; often with donpeacorite	turneaureite, page 233
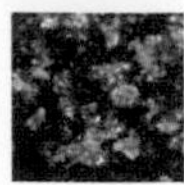	Gray, grainy rock that fluoresces vivid green, from zinc-mining areas	willemite, page 241

A Final Note About Rock and Mineral Identification

When using this book to identify your rock and mineral discoveries, always remember that your specimens can (and likely will) differ greatly than those pictured. Rock and mineral identification isn't easy, and when a specimen is weathered or altered by external forces, it can appear completely different than it "should." So don't necessarily compare your specimen to the photos in this book; learn the key characteristics of each rock or mineral and which traits are constant such as hardness and crystal shape. With a basic understanding of quartz, for example, you'll be able to identify even the most poorly formed specimens.

When it comes to rocks, identification can be very tricky, especially when you're collecting in a state that saw glacial activity. The glaciers of the past ice age pushed enormous amounts of material to places it "shouldn't" be, so it's possible to find odd, perplexing rocks in New York that aren't covered in this book.

Obtaining Permission to Dig

In this book, we mention "obtaining permission" to hunt for minerals in certain areas, particularly the sites of important but now-closed mines. You may wonder how you could go about obtaining permission to dig at these sites. In most cases, tracking down the individual or company who owns a particular mine or quarry may be difficult for the average collector. But local rock clubs often have established histories with site owners and may even take club members on field trips to some of these famous sites. If you can't find a club to join, inquiring at museums with mineral exhibits may get you pointed in the right direction for contacting a site owner. Whatever your solution, if you are fortunate to obtain permission, treat the area and the owner with respect to ensure collecting opportunities in the future.

Samples of pyrite
Limonite coating (brown)
Attached schist
Complex crystal
Cubic crystals
Pyritohedron on dolomite
Pyrite cube (brassy) in schist (dark green)

Sample Page

HARDNESS: 7 **STREAK:** White

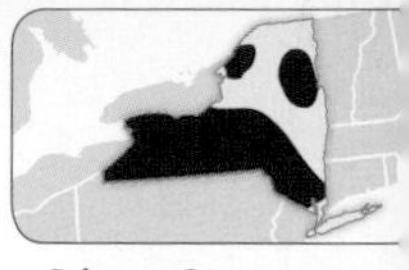
Primary Occurrence

ENVIRONMENT: A generalized indication of the types of places where this rock or mineral can commonly be found. For the purposes of this book, the primary environments listed include **mountains** (particularly in northern New York), **lowlands** (river valleys and plains, particularly in central and western New York), **hills** (most other regions in the state), **rivers** (including riverbanks), **beaches**, **quarries** (and other sites where large amounts of earth have been removed), **outcrops** (exposed bodies of rock), **road cuts** (which are where roads have been cut through hills) and **mines** (including waste-rock piles left over at mine sites).

WHAT TO LOOK FOR: Common and characteristic identifying traits.

SIZE: The general size range of the rock or mineral. The listed sizes apply more to minerals and their crystals than to rocks, which typically form as enormous masses.

COLOR: The colors the rock or mineral commonly exhibits.

OCCURRENCE: The difficulty of finding this rock or mineral. "**Very common**" means the material takes almost no effort to find if you're in the right environment. "**Common**" means the material can be found with little effort. "**Uncommon**" means the material may take a good deal of hunting to find, and most minerals fall in this category. "**Rare**" means the material will take great lengths of research, time, and energy to find, and "**very rare**" means the material is so scarce that you will be lucky to find even a trace of it.

NOTES: These are additional notes about the rock or mineral, including how it forms, how to identify it, how to distinguish it from similar minerals, and interesting facts about it.

WHERE TO LOOK: Here you'll find specific regions or towns where you should begin your search for the rock or mineral.

Fine actinolite crystals

Fibrous mass of intergrown crystals

Actinolite

HARDNESS: 5-6 **STREAK:** White

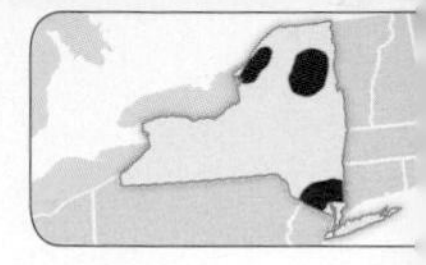
Primary Occurrence

ENVIRONMENT: Mines, quarries, outcrops, road cuts, hills, mountains

WHAT TO LOOK FOR: Elongated, rectangular, fibrous, hard green crystal clusters embedded in metamorphic rocks

SIZE: Crystals can be up to an inch or two; masses and clusters can be up to a foot or more

COLOR: Green to dark green, also gray to brown or nearly black

OCCURRENCE: Uncommon

NOTES: A prominent member of the amphibole group (page 39), actinolite is a very close relative to tremolite (page 229) and is, in essence, the mineral you'd get if you increased tremolite's iron content. As a result, they both share a similar appearance and crystal structure, but actinolite is typically much darker, colored in shades of green. Developed in metamorphic rocks, particularly those rich in magnesium, such as marble or varieties of gneiss, actinolite forms as elongated, generally rectangular crystals with a deeply striated (grooved) or fibrous surface texture, typically in parallel or sometimes radiating clusters. Fine specimens are very translucent, but more abundant are poorly formed, nearly opaque masses. All amphiboles, including actinolite, are most easily confused with pyroxenes (page 181), some of which can also be green, elongated, and just as hard, but can be distinguished by their nearly 90-degree cleavage angles (when pyroxenes are broken, they do so at nearly right angles), while actinolite breaks at steeper angles.

WHERE TO LOOK: Fine specimens have been recovered for years in Orange and Putnam Counties, but the key locality is the Pierrepont area in St. Lawrence County. Less fine crystals are more widespread and found in marble in the southeast.

Examples of amphiboles (dark colored) in rocks

Tremolite crystals

Hornblende (green) in granite

Fibrous "tirodite" (a local name for fibrous cummingtonite and anthophyllite intergrowths)

Amphibole group

HARDNESS: 5-6 **STREAK:** White

Primary Occurrence

ENVIRONMENT: All environments

WHAT TO LOOK FOR: Hard, elongated crystals or grains embedded in coarse-grained rocks; crystals often have a fibrous luster

SIZE: Specimens are rarely larger than an inch or two

COLOR: White to gray or black, light to dark green, brown

OCCURRENCE: Very common

NOTES: The amphiboles are a large group of minerals closely related to the pyroxenes (page 181) and as such are also rock-builders, or minerals most common as constituents of rocks, especially coarse-grained rocks like granite. As a result, most are only seen as small grains or masses identified primarily by their hardness, color, and the typical silky or fibrous structure and luster of many amphiboles. Many amphiboles are present in New York, including the popular tremolite (page 229) and actinolite (page 37), the very common hornblende as well as hastingsite, edenite, cummingtonite, and anthophyllite, to name a few. When it comes to collectible crystals of these minerals, they most often form in pegmatites (very coarse granite outcrops) or several kinds of metamorphic rocks, with the crystals taking several shapes, from elongated prisms to needle-like crystals. Others are extremely fibrous and even flexible—these varieties are asbestos, which is harmful to inhale; see mountain leather (page 167). With so many types, distinguishing amphiboles can be difficult, but it can be done with additional research.

WHERE TO LOOK: The Pierrepont area mines in St. Lawrence County and the Amity-Warwick area of Orange County have produced countless fine specimens. The mountainous areas of Essex County, especially near Keene, yield crystals, too.

Amphibolite
Texture detail
River-worn amphibolite

Amphibolite

HARDNESS: N/A **STREAK:** N/A

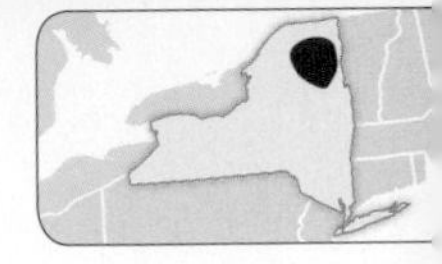
Primary Occurrence

ENVIRONMENT: Mountains, outcrops, rivers

WHAT TO LOOK FOR: Fairly dark, heavy rocks with green and white layering and small-grained texture

SIZE: As a rock, amphibolite can be found in any size

COLOR: Multicolored; light to dark green or yellow-green with white to gray layering, often with brown to black grains

OCCURRENCE: Common

NOTES: Metamorphic rocks can be tricky to identify, especially in the Adirondack Mountains of New York where several types are abundant. Amphibolite is an example of that, as it is often misidentified because it looks similar to other metamorphic rocks and because of its variable appearance. Composed mainly of amphiboles (page 39), a trait for which the rock is named, it forms primarily when diabase (page 57) or gabbro (page 111) are subjected to moderate levels of heat and pressure. The amphiboles and the other minerals within amphibolite, particularly feldspars, are generally granular in nature, not unlike a small-grained sandstone (page 191) in appearance, but the grains in amphibolite are more tightly interlocked. In addition, many (but not all) amphibolites exhibit layering, often with the minerals loosely separated into light and dark layers. In all cases, a predominately light-to-dark green coloration is a key trait, along with a generally mottled, "speckled" look. It can resemble schist (page 195), but is less compressed and splits apart less easily.

WHERE TO LOOK: The Adirondack Mountains and their metamorphic past makes them the only place in the state where amphibolites are easily found. Riverbeds in the vicinity of Lake Placid and northward are ideal for finding samples.

Anorthosite
Weathered specimen
Texture detail
Glassy feldspar grains
Anorthosite

Anorthosite

HARDNESS: ~6-6.5 **STREAK:** N/A

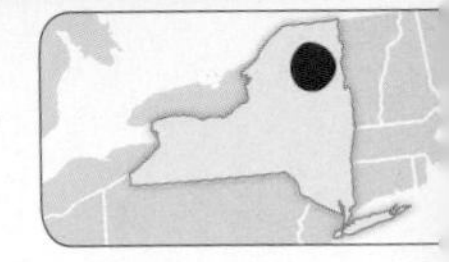

Primary Occurrence

ENVIRONMENT: Mountains, outcrops, road cuts, rivers

WHAT TO LOOK FOR: Coarse-grained rock consisting of almost entirely of glassy gray grains; found in the Adirondacks

SIZE: As a rock, anorthosite can occur in any size

COLOR: Light to dark gray, pale greenish to tan-brown, occasionally pinkish or mottled

OCCURRENCE: Common

NOTES: Anorthosite is an unusual and interesting rock because more than 90 percent of it consists of just one mineral—plagioclase feldspar (page 93) (along with minor amounts of olivine, pyroxenes, and magnetite). Its very coarse, well-crystallized grains formed deep within the earth where its magma cooled very slowly. In fact, because it forms so deeply, anorthosite is rare, as not much of it has been uplifted to the earth's surface. But New York is unusually rich with anorthosite as it makes up much of the core of the Adirondack Mountains, which were dramatically uplifted from the depths and are actually still rising today. Anorthosite is generally light-colored with some translucency, and its coarse feldspar crystals are often rectangular with striations, or fine grooves, running along their length. Some of the region's anorthosite has been metamorphosed to some degree (see page 45), and in all cases can resemble gabbro (page 111), though gabbro is always darker and contains more black mineral grains.

WHERE TO LOOK: Anorthosite makes up much of the Adirondack Mountains; the various peaks in the vicinity of Lake Placid and Saranac Lake are all anorthosite, and any nearby river or stream can contain loose examples.

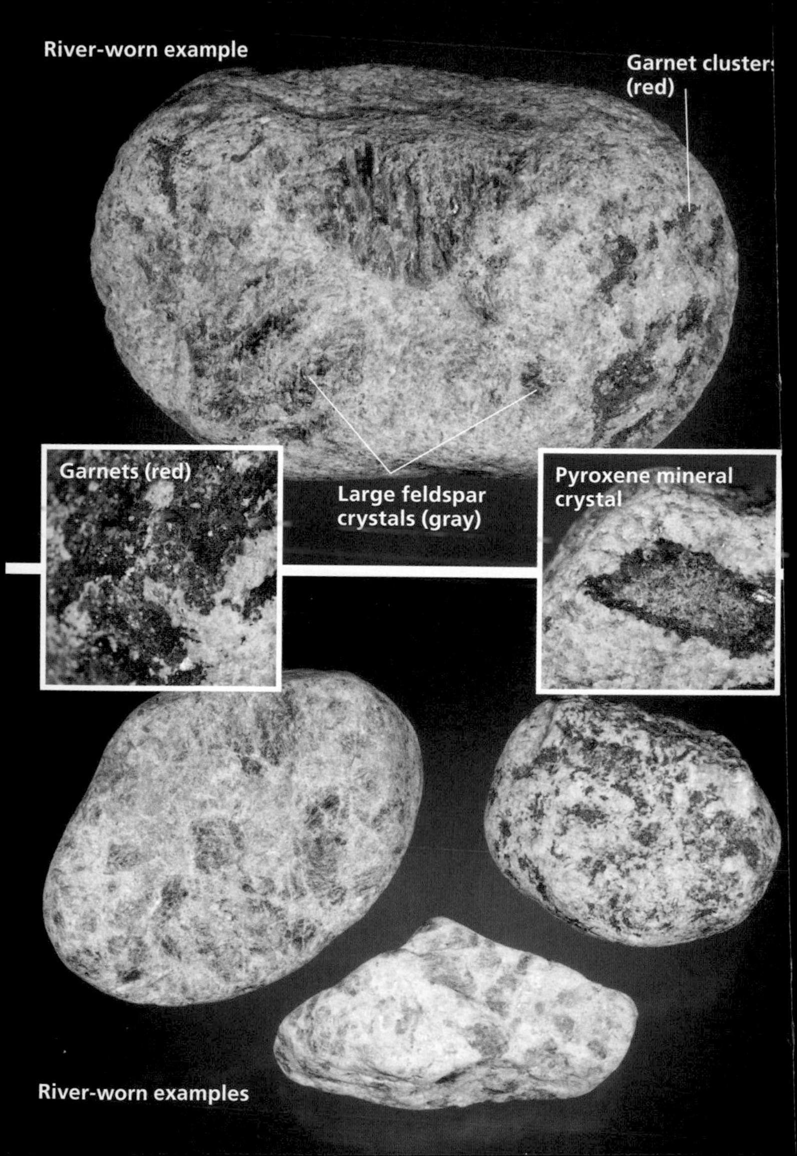
River-worn example
Garnet clusters
(red)
Garnets (red)
Large feldspar
crystals (gray)
Pyroxene mineral
crystal
River-worn examples

Anorthosite, metamorphosed

HARDNESS: ~6-6.5 **STREAK:** N/A

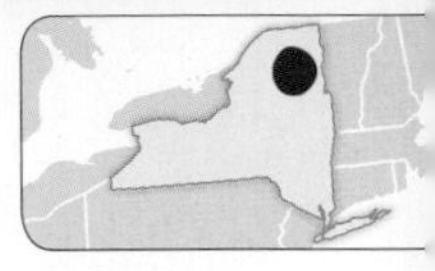

Primary Occurrence

ENVIRONMENT: Mountains, outcrops, road cuts, rivers

WHAT TO LOOK FOR: Light-colored, coarse-grained rock with notably larger embedded crystals and tiny red garnets

SIZE: As a rock, anorthosite can occur in any size

COLOR: Multicolored; predominately gray, with black, bluish, greenish and reddish spots

OCCURRENCE: Common

NOTES: The Adirondack Mountains of New York are an exciting place to study metamorphic rocks, as there are varieties present that would be difficult to find elsewhere in the country. A particularly prominent example is metamorphosed anorthosite, sometimes called meta-anorthosite, which is abundant since the core of the Adirondacks is composed of anorthosite (page 43). As tectonic collisions began mountain-building events throughout the region, heat and pressure on the ancient anorthosite altered its texture and composition. The coarse but more-or-less evenly sized grains of normal anorthosite were altered to a more mixed texture, with large, bluish feldspar crystals embedded in a generally finer-grained mass. Other minerals in the rock were also changed; some into tiny garnets often clustered together as a rim that surround dark pockets of pyroxene minerals. This attractive rock is widespread in the mountains and is easy to find and identify, as the bluish schiller (color flashes) of the feldspars and the bright colors of the ruby-red garnets dotting the rock are distinctive.

WHERE TO LOOK: Metamorphosed anorthosite is only abundant in the Adirondacks, particularly in the central "High Peaks" area. Rivers and outcrops near Lake Placid and Keene produce ample numbers of specimens.

Apatite crystal in pegmatit
Apatite in quartz
Apatite in marble
Mica crystals
Apatite (blue) in marble

Apatite group

HARDNESS: 5 **STREAK:** White

Primary Occurrence

ENVIRONMENT: Mines, quarries, mountains, hills, outcrops, road cuts

WHAT TO LOOK FOR: Hexagonal (six-sided) elongated glassy crystals of moderate hardness, often embedded in quartz

SIZE: Crystals are generally no longer than two inches, and most are less than ½ inch

COLOR: Light to dark blue, green-blue, yellow, rarely colorless

OCCURRENCE: Rare

NOTES: While "apatite" was long considered a single mineral, it is now known to be several chemically distinct minerals that happen to share the same crystal structure, and today they are all considered part of the apatite group. Fluorapatite, the fluorine-bearing apatite variety, is by far the most common type, including in New York, and most unidentified apatite specimens can be assumed to be fluorapatite. Other apatite minerals are present in the state but are far rarer; finding them is unlikely. Fluorapatite is a common constituent of many igneous rocks, but it primarily occurs as tiny inconspicuous grains; only in pegmatites (very coarsely crystallized granite outcrops) and certain metamorphic rocks do crystals grow large enough to collect. Crystals are hexagonal prisms with flat or low-angled tips, are typically translucent and always glassy; many are beautifully colored and embedded in quartz, making for attractive specimens. They can greatly resemble beryl (page 59), but beryl is far harder. Turneaureite is a very rare apatite mineral discussed on page 233.

WHERE TO LOOK: Fluorapatite specimens are found in the rocks of the Adirondacks in Essex County, while St. Lawrence County produces them from pegmatites and skarns (page 209) and Orange County yields specimens in marble.

Clusters of aragonite crystals from sedimentary rock
Aragonite mass
Crystals embedded in mass
Same specimen as above under shortwave ultraviolet light

Aragonite

HARDNESS: 3.5-4 **STREAK:** White

Primary Occurrence

ENVIRONMENT: Lowlands, mines, quarries, outcrops

WHAT TO LOOK FOR: Light-colored masses or needle-like crystals similar to calcite, but harder

SIZE: Crystals are usually under ¼ inch, while masses may be up to fist-sized

COLOR: Colorless to white, gray, yellow to brown, pinkish

OCCURRENCE: Uncommon

NOTES: A close cousin to calcite (page 63) and an even closer cousin to strontianite (page 217) and cerussite (page 69), aragonite is a calcium carbonate that forms in a number of geological environments, yet collectible specimens aren't very common in New York. Despite bearing the same chemical composition as calcite, the two are distinct minerals due to differing crystal structures; calcite generally forms as tapered hexagonal points or rhombohedrons (shapes like a leaning cube), aragonite develops as slender, needle-like crystals or fibrous masses. Aragonite is also harder than calcite, which is a key distinction. Massive varieties are more easily mistaken for gypsum (page 125), but gypsum is also notably softer than aragonite. Aragonite is found in limestone deposits as crystals, veins, or masses, but fine crystals are also found in metamorphic rocks as well as in metallic ore-bearing regions. In all cases, aragonite is a low-pressure mineral, which means it doesn't take a mine to reveal specimens, and they instead can be found in much more accessible areas.

WHERE TO LOOK: The Tilly Foster Iron Mine in Putnam County produces very fine specimens, but is off-limits to those without permission. Elsewhere, aragonite is found sparingly in limestone formations as well as in metamorphic rocks.

Arsenopyrite (silvery) in quartz
Tiny ⅛" crystal
Embedded crystals
Crystals embedded in quartz

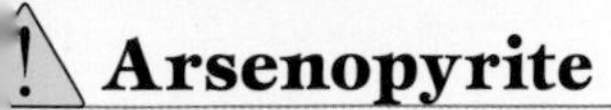

Arsenopyrite

HARDNESS: 5.5-6 **STREAK:** Grayish black

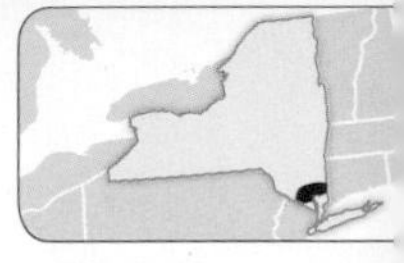

Primary Occurrence

ENVIRONMENT: Mines, quarries, outcrops

WHAT TO LOOK FOR: Fairly hard, metallic silvery masses or rectangular crystals embedded in rock, particularly marble

SIZE: Specimens are typically smaller than a few inches

COLOR: Steel-gray to silvery white; can also have a brassy coating

OCCURRENCE: Rare

NOTES: As its name suggests, arsenopyrite consists of iron, sulfur, and arsenic, and it is a rare New York mineral associated primarily with the metamorphic rocks in the southeastern corner of the state. Occurring most often in the area's marble (page 157), arsenopyrite appears as lustrous, silvery grains or masses embedded in the stone, often without much crystal shape. Well-formed crystals, though extremely rare, take the form of steeply pointed diamond-like shapes, which are sometimes flat and plate-like, but also occur in thicker, blockier pieces, and all crystals tend to exhibit deeply striated (grooved) surfaces. This dense and brittle mineral could be confused with pyrite (page 177), particularly when weathered, but pyrite is far more common throughout the state. It is more likely that arsenopyrite could be confused with löllingite, a rare mineral with which it can occur; löllingite has a nearly identical appearance, luster, hardnesses, and streaks, but its crystals are more elongated. If found in grains or masses, there is no practical way an amateur can distinguish them. Assume you've found arsenopyrite, however, as it is a bit more common than löllingite.

WHERE TO LOOK: Most of New York's localities are in Orange and Putnam Counties; it's still possible to find some small pieces in the old mine dumps or outcrops near Warwick and Amity.

Mat of needle-like crystals on serpentine

Crystal detail

Clusters of tiny ⅛" crystals on serpentine

Artinite

HARDNESS: 2.5 **STREAK:** White

Primary Occurrence

ENVIRONMENT: Mines, outcrops

WHAT TO LOOK FOR: "Fuzzy" groupings of tiny, white, needle-like crystals on serpentine

SIZE: Crystal mats can be up to palm-sized and rarely larger

COLOR: White, rarely colorless

OCCURRENCE: Very rare; no longer possible to collect

NOTES: There are several New York minerals that are found at only a single site, but artinite is perhaps the best known of these rarities. Appearing as delicate, "fuzzy" mats or crusts of countless tiny needle-like crystals, New York's artinite is only found on Staten Island where collecting sites were briefly uncovered during construction projects and then buried once again. It forms as an alteration product of serpentines (page 199), forming when serpentines are affected chemically. It therefore is found in cavities and fractures or veins within serpentine formations. Because of its unique appearance and close association with serpentine minerals, there is only one mineral with which artinite can be easily confused: hydromagnesite (page 139). Hydromagnesite occurs in the same environment and often on the same specimens as artinite, and while both form tiny delicate needles, hydromagnesite's tend to be much shorter and more poorly developed, appearing as a chalky crust rather than a mat of "fuzz."

WHERE TO LOOK: New York's best known locality is the Spring Street Occurrence, a serpentine outcrop on Staten Island. Unfortunately, though that outcrop is still present today as a small hill, it is private, surrounded by fencing and off-limits to collectors. A rock shop may be your only option to obtain a specimen.

Cluster of translucent baryte crystals
Wedge-shaped tip
Blocky crystals
Tightly intergrown bladed crystals on larger baryte mass

Baryte (Barite)

HARDNESS: 3-3.5 **STREAK:** White

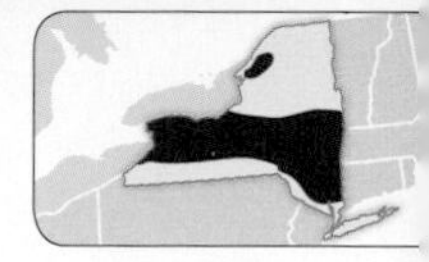

Primary Occurrence

ENVIRONMENT: Lowlands, hills, mountains, quarries, mines, outcrops, road cuts

WHAT TO LOOK FOR: Light-colored blocky or bladed crystals that feel heavy for their size

SIZE: Crystals are typically thumbnail-sized or smaller

COLOR: Colorless to white or gray, tan to brown, reddish

OCCURRENCE: Common

NOTES: Baryte (often spelled "barite," but only in the U.S.) is the most abundant barium-bearing mineral and is common, especially in sedimentary rock environments where it can often be found well crystallized in pockets within limestone. This popular mineral is generally light-colored and glassy, forming blocky, angular or blade-like crystals with wedge-shaped tips, and crystals are typically intergrown in more-or-less parallel clusters. Because of its lower hardness it can be confused with other similar soft minerals, but it can be differentiated by its high specific gravity, or density, which makes even a small specimen feel heavy for its size, an unusual trait for a glassy mineral. Poorly crystallized specimens can resemble calcite (page 63) or fluorite (page 95), but baryte is harder than calcite and softer than fluorite. It is easily confused with celestine (page 67), which is rarer and often blue. Aside from sedimentary rocks, baryte is also abundant in metallic ore mines, such as the zinc mines of northern New York, where it occurs with sphalerite (page 213).

WHERE TO LOOK: St. Lawrence County has produced more baryte (including high-quality specimens) than any other area. Macomb, Gouverneur, and Balmat-Edwards areas have historically been lucrative. Any limestone in the southern half of the state also has potential such as in Monroe County.

Basalt
Waterworn surface
Brown weathered surfaces
Diabase texture detail
Diabase
Coarser grained texture than basalt

Basalt/Diabase

HARDNESS: 5-6 **STREAK:** N/A

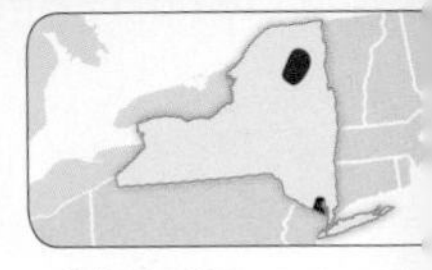

Primary Occurrence

ENVIRONMENT: Hills, outcrops, road cuts, rivers

WHAT TO LOOK FOR: Dark gray to black, fine-grained, hard, dense rocks, sometimes with vesicles (gas bubbles)

SIZE: As rocks, basalt and diabase can be found in any size

COLOR: Gray to black, sometimes with a greenish tint or mottling; browner when weathered

OCCURRENCE: Uncommon to rare, depending on region

NOTES: Basalt and diabase are prevalent all over the planet and form when volcanic eruptions release lava (molten rock) on or near the earth's surface. Both rocks consist of essentially the same minerals—plagioclase feldspar, pyroxenes, amphiboles, olivines, and magnetite—but they are considered different rocks because of the varying sizes of their mineral grains. Basalt results when the magnesium- and iron-rich lava from which it forms is erupted onto the surface where the earth's atmosphere causes it to cool and solidify very rapidly, allowing no time for its minerals to form visible crystals and often trapping gas bubbles to form cavities (called vesicles). Diabase, on the other hand, forms near the surface in settings where it cools fairly quickly, but more slowly than basalt, causing larger mineral grains to form. Telling the two apart can be tricky, as they're both dark, dense, and occasionally weakly magnetic, but in diabase it shouldn't be difficult to find small, rectangular, glassy feldspar crystals, and diabase contains no vesicles. Gabbro (page 111) is also closely related, containing the same minerals, but instead formed deeply within the earth.

WHERE TO LOOK: The glaciers deposited pebbles all over, but the only notable basalt formation is in Rockland County. The Hudson River will reveal samples. Diabase formed sparingly in the Adirondacks, and can be recovered from rivers there.

Poorly formed greenish mass from pegmatite
Hexagonal cross section
Yellow crystal
Mica crystals
Large yellow crystal on pegmatite

Beryl

HARDNESS: 7.5-8 **STREAK:** Colorless

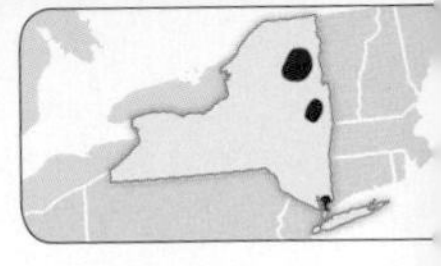
Primary Occurrence

ENVIRONMENT: Mines, quarries, hills, outcrops

WHAT TO LOOK FOR: Very hard, elongated hexagonal (six-sided) crystals embedded in coarse-grained rocks

SIZE: Crystals and masses are often an inch or more in size; extremely rare specimens have measured up to a foot

COLOR: Gray, yellow to greenish, rarely blue

OCCURRENCE: Rare

NOTES: Beryl is a long-cherished mineral perhaps better known by the names given to its more colorful varieties, emerald and aquamarine. It is a rare beryllium- and aluminum-bearing silicate that most commonly forms in pegmatites, which are extremely coarse-grained granite formations that solidified deep within the earth where its minerals could crystallize to very large, collectible sizes. In this environment, beryl is typically found tightly embedded and hidden in feldspars, quartz, and micas, but its glassy crystals are generally so distinct in shape that they can still easily be identified. It forms as perfect hexagonal (six-sided) prisms with flattened tips and occasionally develops in columnar (parallel) groups. The best specimens are translucent, but most are nearly opaque, especially when fractured or otherwise poorly formed. Massive specimens lacking beryl's distinctive shape are still easy to identify, thanks to its very high hardness and typical colors. Chrysoberyl (page 79) is easily confused with beryl, but is even harder and rarer, and apatite group minerals (page 47) can look very similar, but all are much softer.

WHERE TO LOOK: The Bedford area in Westchester County is home to a significant pegmatite outcrop, and specimens may still be collected today from exposed rock. Specimens have also been obtained in Saratoga County quarries.

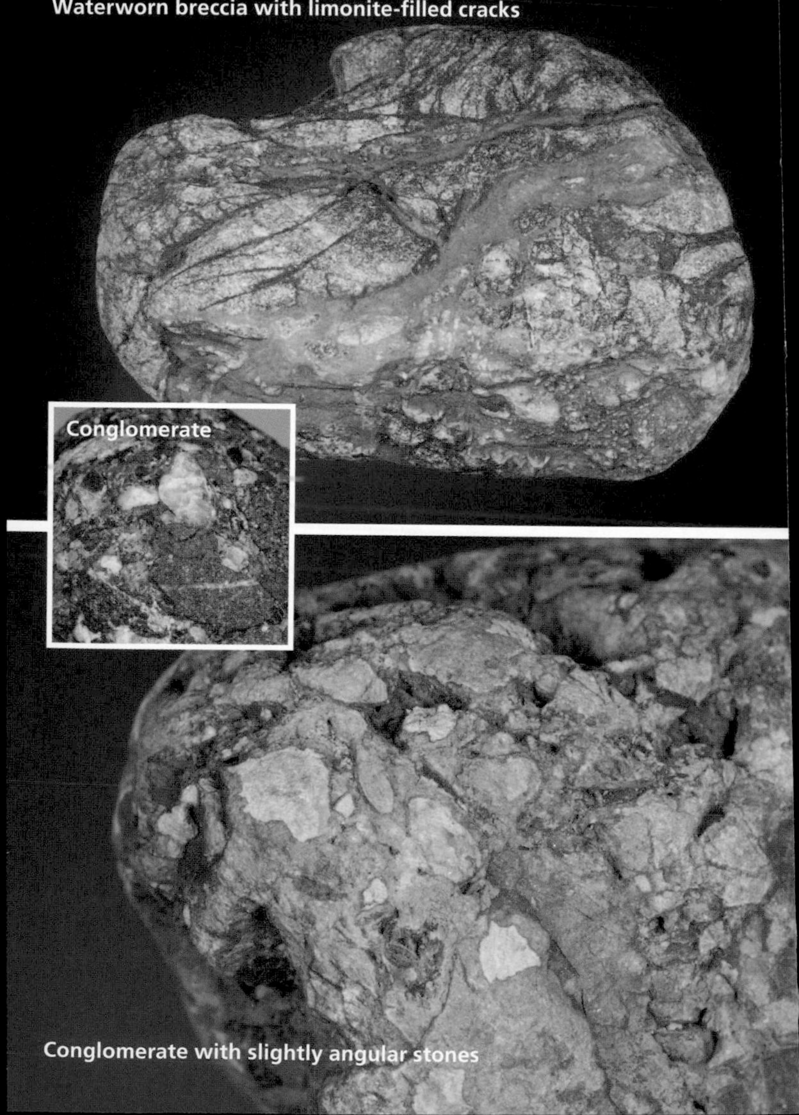
Waterworn breccia with limonite-filled cracks
Conglomerate
Conglomerate with slightly angular stones

Breccia/Conglomerate

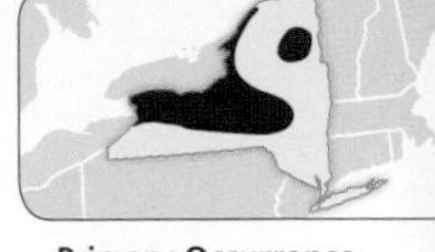

Primary Occurrence

HARDNESS: N/A **STREAK:** N/A

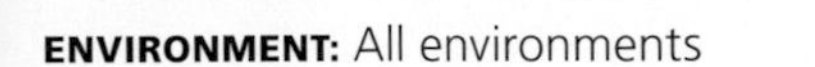

ENVIRONMENT: All environments

WHAT TO LOOK FOR: Rocks that appear to be made of smaller rocks or rock fragments cemented together

SIZE: As rocks, breccia and conglomerate can be found in any size

COLOR: Varies greatly; mottled and multicolored

OCCURRENCE: Common

NOTES: Breccia and conglomerate are two similar sedimentary rocks that formed when smaller rocks or rock fragments were solidified together by a finer-grained cementing sediment, giving them distinctly uneven textures. The rocks within breccia and conglomerate can be virtually any other type of rock, depending on what was present when it formed. Breccia and conglomerate are considered different rocks because the smaller stones that make them up are shaped differently. Breccia is formed of the fractured, angular fragments of a preexisting rock that was broken and crushed in a geological event such as the pressure of metamorphism or during a volcanic eruption; conglomerate formed amid aquatic conditions when rounded, waterworn pebbles were locked together by finer sediment. In each case, the cementing material is often material like sand, clay, or limonite (page 151), or even minerals like calcite and quartz. Identifying these rocks is as easy as noting their distinct appearance and texture. Waterworn concrete may initially appear to be conglomerate but is generally much harder.

WHERE TO LOOK: Any area that saw lots of metamorphism, such as the Adirondacks, should yield breccia specimens in rivers and outcrops. Lowland areas rich with sedimentary rocks will also bear conglomerates, particularly along rivers.

Very fine calcite crystal from limestone pocket
Rhombohedral calcite cleavage
Calcite vein in limestone
Very fine calcite crystal cluster from limestone

Calcite

HARDNESS: 3 **STREAK:** White

Primary Occurrence

ENVIRONMENT: All environments

WHAT TO LOOK FOR: Light-colored pointed crystals or blocky masses or veins that can be easily scratched with a U.S. nickel

SIZE: Crystals can measure anywhere from a fraction of an inch up to several inches; masses can be many inches

COLOR: Colorless to white; often yellow to brown or honey-color

OCCURRENCE: Very common

NOTES: Like quartz (page 185), calcite is so common that it is one of the first minerals all collectors should learn to recognize and identify. Composed of calcium carbonate, calcite can form in virtually any geological environment and as a result has several hundred known crystal forms. Most commonly, however, it develops as steeply pointed, six-sided crystals, or as rhombohedrons (a shape resembling a leaning cube), though more complex shapes are common as well; no matter its form, well-developed crystals are typically glassy and translucent. But irregular masses, crusts and veins embedded in rocks are more abundant, and are often opaque. But in any form, calcite is easy to identify. It is soft enough to be scratched with a U.S. nickel, which will distinguish it from similar-looking quartz. It will also effervesce, or fizz, when exposed to even weak acids such as undiluted vinegar, due to calcite's carbonate content. Lastly, calcite has perfect rhombohedral cleavage, so when struck, it will break into rhombohedrons. Dolomite, baryte, fluorite, celestine, and aragonite can all appear similar, but they all are harder.

WHERE TO LOOK: Calcite is ubiquitous and can be found in some form virtually anywhere. The southern and western portions of the state are particularly known for fine crystals, especially in Herkimer, Niagara Falls, and in Rochester quarries.

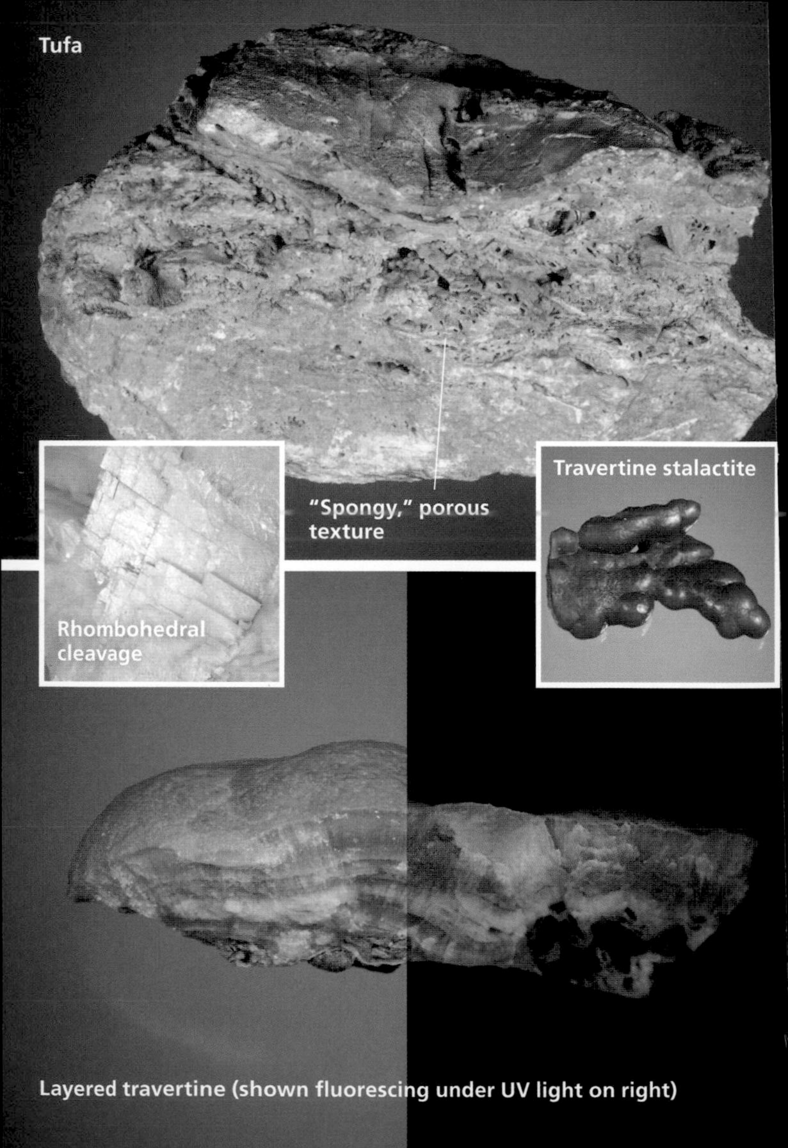
Tufa
"Spongy," porous texture
Travertine stalactite
Rhombohedral cleavage
Layered travertine (shown fluorescing under UV light on right)

Calcite, varieties

HARDNESS: 3 **STREAK:** White

Primary Occurrence

ENVIRONMENT: All environments; particularly lowlands, rivers, quarries

WHAT TO LOOK FOR: Masses or coatings of soft material with all the traits of calcite

SIZE: Calcite varieties can range up to several feet in size

COLOR: Colorless to white, yellow to tan or brown very common

OCCURRENCE: Very common

NOTES: Calcite is abundant in so many environments that many varieties, both collectible and mundane, exist throughout the state. Limestone (page 149) is a sedimentary rock that consists largely of calcite; it is not a variety of calcite, but is typically the host for calcite and its varieties such as tufa. Tufa is an opaque, chalky, calcite-rich material formed when surface water deposits dissolved calcite, often with large amounts of organic matter. The organic material gives the deposit a "spongy" texture and forms a light, cavity-filled rock often loaded with fossils. But travertine is a far more attractive and collectible variety of calcite; travertine is layered calcite formation that developed when dissolved calcite was deposited by heated groundwater. In New York, travertine is often translucent brown or honey-colored with interesting bulbous or stalactitic (icicle-like) shapes and waxy surfaces, which make for interesting, sculptural specimens. When fairly pure, most calcite varieties share the identifying traits of typical calcite; impure varieties, like tufa, however, are more difficult to identify, though they should still effervesce in acids, just like pure calcite does.

WHERE TO LOOK: Ilion Gorge in Herkimer County is well known for its fine specimens of travertine that form at fractures in the cliff walls; area rock clubs make frequent field trips here.

Cerussite (white/tan) on large cubic galena crystal

Cerussite

HARDNESS: 3-3.5 **STREAK:** White

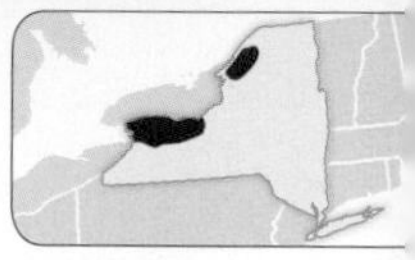

Primary Occurrence

ENVIRONMENT: Lowlands, hills, quarries, road cuts, mines

WHAT TO LOOK FOR: Light-colored, dull, dusty crusts or coatings on galena

SIZE: Crusts may be several inches; crystals are rare and tiny, smaller than ⅛ inch

COLOR: Most commonly white to gray or tan to brown

OCCURRENCE: Common

NOTES: Easily overlooked and rarely exciting, cerussite is a New York mineral that isn't typically worthy of mention. But this typically mundane material is abundant enough that collectors in the state should be able to identify it. Related to aragonite (page 49), cerussite is a lead carbonate (generally safe, but wash your hands after handling) that forms primarily as a weathering product of galena (page 113). As such, it is virtually always found in the presence of galena and most often on the surface of galena crystals. In this setting, it appears as dull, often chalky or dusty crusts or thin coatings on the galena, dimming the otherwise brightly lustrous, metallic appearance of the galena. Crystals are very rare in New York, but appear as tiny, glassy, needle-like or ball-like crystals; purer specimens are whiter in color while the majority of samples are gray or tan. You won't likely confuse cerussite's lackluster crusts with anything else, because its association with galena and other lead-bearing minerals is highly distinctive. Some amount of cerussite can be assumed present with all specimens of galena.

WHERE TO LOOK: Where you find galena, you'll find cerussite. The Balmat-Edwards and Rossie areas in St. Lawrence County produce many specimens.

Attached baryte crysta
Waxy surfaces
Irregularly shaped chalcedony mass from a pocket in sedimentary rock
River-worn samples deposited by glaciers

Chalcedony

HARDNESS: ~7 **STREAK:** N/A

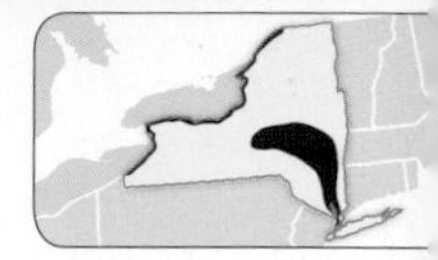

Primary Occurrence

ENVIRONMENT: All environments

WHAT TO LOOK FOR: Very hard, often waxy masses of translucent material, generally with uneven coloration

SIZE: Specimens of chalcedony are typically smaller than a fist

COLOR: White to gray if pure, but more commonly brown to reddish, yellow

OCCURRENCE: Common

NOTES: Like jasper (page 143) and chert (page 77), chalcedony is a microcrystalline form of quartz (page 185), which means that it consists of densely packed, intergrown microscopic quartz crystals and has no regular outward shape. Instead, its shape is governed by its surroundings. Chalcedony is best known as the material that agates are made of; agates are famous for their concentric rings of differently colored chalcedony, but they are quite rare in New York. Because it is a variety of quartz, chalcedony shares quartz's high hardness, translucency, resistance to weathering, and conchoidal fracture (when struck, circular cracks appear), all of which will help identify it. Chert and jasper share many of the same features, but are far more opaque due to their more granular microstructure as opposed to chalcedony's more fibrous microstructure. Chalcedony is often found in gravel, particularly in rivers, as rounded, waxy masses, but other growths of it, such as botryoidal (grape-like) crusts or odd free-standing shapes, may be found in cavities in rocks, often with a coating of drusy quartz.

WHERE TO LOOK: Much of New York's chalcedony was deposited by glaciers and therefore can be found throughout the state in gravel and rivers. Cavities in limestone in central New York may yield crusts or masses of chalcedony as well.

Chalcodite crust (metallic brown and brassy colored)
Radial crystal cluster

Tiny (1⁄16") chalcodite blades (metallic brown) on reddish quartz

Chalcodite

HARDNESS: 1 **STREAK:** Yellow to brown

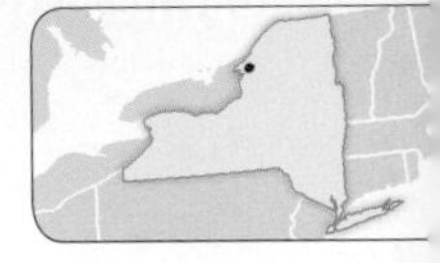

Primary Occurrence

ENVIRONMENT: Mines

WHAT TO LOOK FOR: Very soft, metallic radiating clusters of tiny plate-like crystals

SIZE: Clusters are tiny, smaller than a pea

COLOR: Bronze-brown to brassy

OCCURRENCE: Very rare

NOTES: Chalcodite is not generally considered to be a distinct mineral, but instead is the rarer, more iron-rich variety of the mineral stilpnomelane. Stilpnomelane itself is an interesting mineral that forms as black, plate-like crystals embedded in metamorphic rocks, but it is virtually absent from the state. Chalcodite isn't much more common, however; in New York it is only found in a single mine in Jefferson County, though the mine's specimens are of a fantastic quality. It appears as extremely soft, metallic brown or bronze-colored flakes or plate-like crystals, always tiny and often clustered together in radiating, rounded, ball-like formations. Samples generally occur on quartz (page 185) with siderite (page 203) within cavities in the red, hematite-rich rock matrix of the mine, making for attractive specimens, especially when viewed under magnification. Chalcodite specimens can resemble mica group minerals (page 161), but since no micas occur in the mine, any possibility for confusion is eliminated.

WHERE TO LOOK: In New York, only the Sterling Mine in Antwerp produced this rare variety of stilpnomelane, and the area is off-limits to those without permission. Luckily, specimens are still available on the market, however, due to the popularity of the mine's mineral specimens.

River-worn chert pebbles

Flint (black chert)

Waxy surfaces

Chert-filled fossil in limestone

Chert

HARDNESS: ~7 **STREAK:** N/A

Primary Occurrence

ENVIRONMENT: All environments

WHAT TO LOOK FOR: Very hard, dark, opaque masses with a waxy texture and appearance on worn or broken surfaces

SIZE: Chert can occur in virtually any size but is often found as masses smaller than a fist

COLOR: White to gray if pure, but more often brown to black

OCCURRENCE: Very common

NOTES: Chert is a very hard, dense and opaque sedimentary rock that consists almost entirely of tightly bonded microscopic grains of quartz (page 185). It can form in several ways and in a number of different geological environments, but most of New York's chert is either found as masses in limestone that formed when silica (quartz material) collected in cavities left behind by dissolved fossil material, or as large layered beds that developed when the skeletons of diatoms (nearly microscopic algae that grows rigid skeletons made of silica) settled onto seafloors in thick beds that later solidified. Chert is typically one of the hardest rocks you'll encounter but is matched by chalcedony (page 71), which is far more translucent, and quartzite (page 189), which is typically grainier and glassier in appearance when broken. Chert is often dull, but is so hard that weathering can nearly polish it, resulting in smooth, waxy surfaces when water-worn pebbles of it are found. Chert also shares quartz's conchoidal fracture (when struck, circular cracks appear), which is an identifying trait.

WHERE TO LOOK: Widespread and easy to identify, chert is abundant in most of New York's sedimentary areas. Rivers in the Catskills will yield freed examples, as will Finger Lakes beaches, and specimens still embedded in their host limestone may be found at any outcrop.

Chrysoberyl mass (pale yellow-green) in pegmatite

Pale-green mass

Crude chrysoberyl crystal fragment on quartz in pegmatite

Chrysoberyl

HARDNESS: 8.5 **STREAK:** White

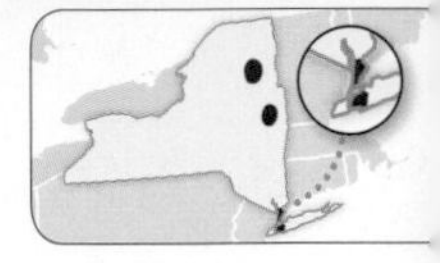
Primary Occurrence

ENVIRONMENT: Hills, quarries, mines, outcrops

WHAT TO LOOK FOR: Extremely hard, very scarce, light-colored glassy masses or crystals embedded in pegmatite

SIZE: Crystals are up to an inch or two, while masses may rarely be up to fist-sized

COLOR: Light to dark green, yellow-green, pale to dark yellow

OCCURRENCE: Very rare

NOTES: A very scarce beryllium-bearing mineral, chrysoberyl has long been used as a gemstone. Forming exclusively embedded within granite pegmatites (very coarsely crystallized granite outcrops) it is closely associated with quartz, micas, feldspars, and tourmalines. Chrysoberyl isn't often found finely crystallized, but when it is, it takes the form of tabular (flattened) or short prismatic crystals with striated (grooved) faces. Generally translucent and greenish yellow in color, these crystals are frequently twinned, or intergrown within each other at specific angles. Some twinned specimens are notably heart-shaped, which make for striking specimens. Irregular masses are far more common, however, and these often fill spaces between other crystals in the pegmatite, and may be whiter, more opaque, or otherwise not ideally formed. However it may appear, it isn't hard to identify due to its extremely high hardness, which distinguishes it from the quartz and feldspars it occurs with as well as from beryl (page 59), which is the most likely source of confusion.

WHERE TO LOOK: The Greenfield area of Saratoga County has produced the state's best specimens from pegmatite quarries. Ironically, some fine specimens have originated in the rock beneath New York City, which are, of course, completely unobtainable.

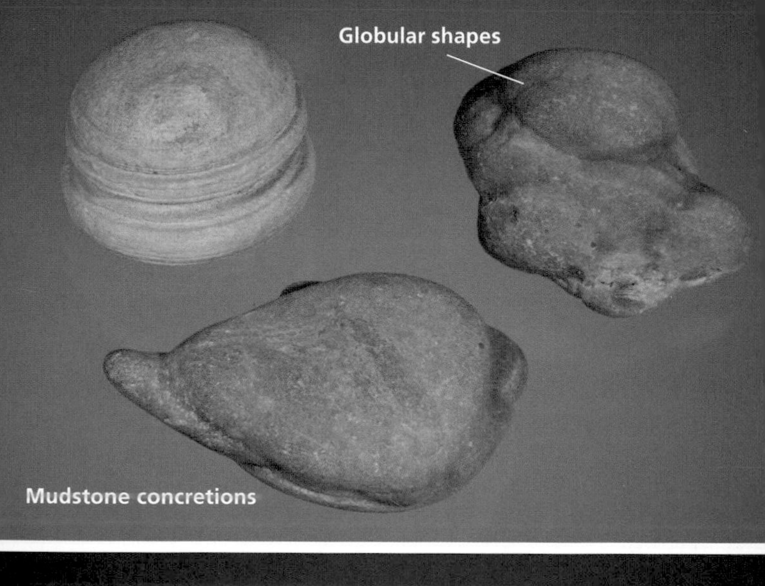
Globular shapes
Mudstone concretions

Conspicuously round shape
Mudstone concretions

Concretions

HARDNESS: N/A **STREAK:** N/A

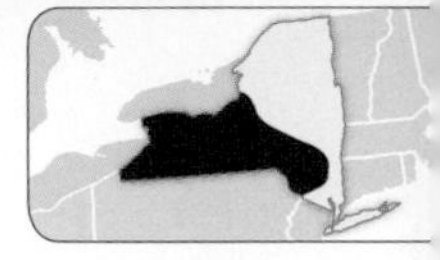

Primary Occurrence

ENVIRONMENT: Lowlands, quarries, outcrops, rivers, road cuts

WHAT TO LOOK FOR: Conspicuously round rocks found embedded within or weathered free from other sedimentary rocks

SIZE: Concretions are usually smaller than a fist, but may be up to a foot or more

COLOR: Gray to tan or brown, yellowish; often depends on host rock

OCCURRENCE: Uncommon

NOTES: A concretion is a round, ovoid, or globular body of sedimentary rock that has gradually concreted around a nucleus, or central point. Often, the nucleus is a fragment of fossil organic material that released carbon as it decayed; this carbon reacted with the surrounding rock and formed a mineral "rind" around the fossil. Concretions develop within other sedimentary rocks, including shale (page 201), mudstone (page 169), and sandstone (page 191), and can be found still embedded in those rocks as well as in specimens that have weathered free. When found in host rock, identification is easy, as their oddly round shapes are conspicuous and they are often harder than the surrounding rock. Those that are weathered free, however, may resemble any other rounded, worn rock. Identifying these specimens is trickier, but it may help to determine if its texture is still grainy or rough enough; this can help rule out look-alikes that were actually water-worn and tend to be very smooth.

WHERE TO LOOK: Any low-lying sedimentary region has the potential to produce concretions. The shales of the Finger Lakes area rivers and lakes are rich with fossils as well as concretionary formations. The Catskills are lucrative as well.

Datolite crystals (white) with calcite (darker tan/brown)

Lustrous crystals

Fine crystals

Complex, angular datolite crystals in cavity

Datolite

HARDNESS: 5-5.5 **STREAK:** White

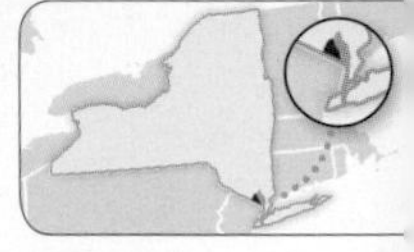

Primary Occurrence

ENVIRONMENT: Mines, quarries, outcrops, hills

WHAT TO LOOK FOR: Brightly lustrous, glassy, very small crystals with a flattened, complex shape

SIZE: Crystals are very small, less than ¼ inch

COLOR: Colorless to white, sometimes with a greenish tinge

OCCURRENCE: Very rare

NOTES: Found in several environments but particularly within cavities in igneous rocks like basalt (page 57) and diabase (page 57), or in metamorphic rocks like skarn (page 209, datolite is a very rare find in New York, with only a handful of localities having ever produced it. A calcium- and boron-bearing silicate mineral, datolite develops in the region as small, brightly lustrous, generally colorless crystals with complex, multi-faceted shapes, that are sometimes ball-like in appearance but more commonly slightly flattened with two faces more broad than the others. The shape of fine crystals may be difficult to observe, however, due to their small size and because they frequently intergrow in complex clusters, often with similarly colored calcite (page 63) which can hinder identification. It can also occur with zeolites (page 245), some of which can share a vaguely similar shape, but datolite is harder than either zeolites or calcite. Datolite can also easily be mistaken for quartz (page 185), but quartz is much harder.

WHERE TO LOOK: Most localities for New York datolite are in Rockland County such as in quarries near Nyack and Orangetown. But these heavily populated area means that most sites are private or hidden beneath urban construction.

Extremely fine, translucent diopside crystal from De Kalb

Crystals in marble

Diopside

HARDNESS: 5.5-6.5 **STREAK:** White to gray

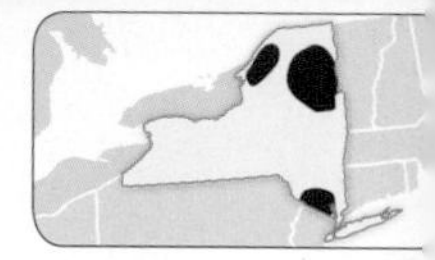

Primary Occurrence

ENVIRONMENT: Mines, quarries, road cuts, outcrops, mountains, hills

WHAT TO LOOK FOR: Fairly hard, greenish, blocky crystals with square cross sections in metamorphic rocks

SIZE: Crystals are generally thumbnail-sized or smaller

COLOR: Colorless to white or gray, yellowish, pale to dark green

OCCURRENCE: Uncommon

NOTES: One of the most prominent pyroxene group (page 181) minerals in New York State and certainly the most collected, diopside is a fairly abundant calcium- and magnesium-bearing silicate mineral found primarily in metamorphic rocks. And since New York is home to countless formations of gneiss (page 117), marble (page 157), and skarn (page 209), all metamorphic rocks in which diopside is most often found, there are dozens of prolific collecting sites to visit. When well-crystallized, diopside forms blocky prismatic crystals with angled tips and almost perfectly square cross sections, and it is frequently found alongside garnets, humite group minerals, and feldspars. Nonetheless, it is more abundant as indistinct grains or small masses embedded in rock; this is its typical appearance when found in marble. It is generally pale to dark green in color, glassy, and fairly hard, but when crystals aren't present, it can be difficult to spot and identify. Many specimens are fluorescent pale blue under shortwave UV light, which may help. It can resemble olivine group minerals (page 171), but diopside is softer.

WHERE TO LOOK: The De Kalb area of St. Lawrence County once produced some of the finest gem-grade diopside ever found in the U.S. Mountainous areas of Essex County still produce embedded specimens at road cuts and outcrops.

Very fine dolomite crystal cluster from limestone pocket

Marcasite crysta

Fine crystal cluster

Curved, rhombohedral dolomite crystals (tan) on marcasite (brassy)

Dolomite

HARDNESS: 3.5-4 **STREAK:** White

Primary Occurrence

ENVIRONMENT: Lowlands, hills, quarries, outcrops, road cuts, mines

WHAT TO LOOK FOR: Blocky, rhombohedral (shaped like a leaning cube) light-colored crystals with pearly, curved surfaces

SIZE: Single crystals are rarely larger than a thumbnail

COLOR: White to gray, pinkish to tan, brown; less commonly yellow to orange or reddish

OCCURRENCE: Very common

NOTES: A common mineral most often found in cavities within sedimentary rocks or as a constituent of the rocks themselves (primarily limestone), dolomite is a close cousin to calcite (page 63). It is also a mineral with which all collectors should be familiar. Like calcite, dolomite develops as rhombohedral crystals, which resemble a leaning or skewed cube, but it also often exhibits rounded or slightly curved crystal faces; these crystal faces have a pearly luster and softly reflect light. This distinguishes it from calcite in most well-crystallized specimens, but hardness is always a fail-safe identifier, as dolomite is harder. Siderite (page 203) can also appear very similar and is of a similar hardness, but typically occurs with iron-bearing minerals like hematite (page 129), and it is less common. Sharp crystals of dolomite are highly collectible, especially when alongside other minerals such as marcasite and sphalerite. Finally, a variety of dolomite-rich limestone, called dolostone, is prevalent, but it is visually identical to limestone.

WHERE TO LOOK: Wayne, Niagara, Monroe, and Herkimer Counties are all known for excellent, sharp crystals lining cavities in limestone or dolostone. The famous pay-to-dig Herkimer area "diamond" (quartz) mines produce fine specimens, as do quarries near Rochester.

Cluster of intergrown donpeacorite crystals

Orange mass in matrix

Intergrown crystals

Crust of flat, striated (grooved) tan to orange crystals

Donpeacorite

HARDNESS: 5-6 **STREAK:** White

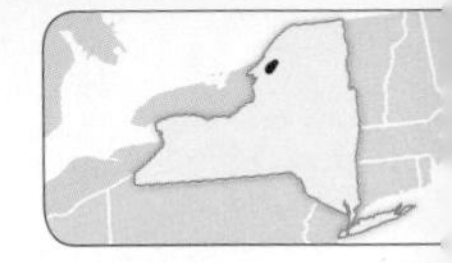

Primary Occurrence

ENVIRONMENT: Mines

WHAT TO LOOK FOR: Minute, flat, light-colored crystals embedded in rock from the Balmat-Edwards area

SIZE: Crystals remain smaller than ¼ inch, but may be intergrown in masses measuring an inch or two

COLOR: Pale yellow to tan, pale orange, faint pink

OCCURRENCE: Very rare

NOTES: An extremely rare member of the pyroxene group (page 181), donpeacorite is currently found at only three sites on earth, with the ZCA No. 4 Mine in the Balmat-Edwards mining district of St. Lawrence County being the original site of discovery. A relatively "new" mineral, it was discovered in 1984 and only occurs as tiny, flat crystals embedded in rock from the mine, and its glassy, light-colored, generally tan crystals can almost exactly match the color of their host rock, making them very easy to overlook. When large enough to observe, its crystals tend to exhibit a slightly fibrous appearance, but that may only serve to cause further confusion as it occurs with cummingtonite, an amphibole (page 39) that is also fibrous. Since donpeacorite only formed in rock from a certain depth of the mine, not all mine material will contain it, so using turneaureite (page 233) and its trademark fluorescence as an indicator will help, as the two are often found in the same specimen. Distinguishing donpeacorite from cummingtonite is more difficult, but generally donpeacorite is darker in color.

WHERE TO LOOK: The ZCA No. 4 Mine in the Balmat-Edwards area is not only the single source in New York State, but in all of North America. But the mine is long closed; realistically, purchasing a specimen may be your only option.

Tiny epidote crystals (light to dark green) on matrix

Mass of epidote from a vein in igneous rock

Epidote

HARDNESS: 6–7 **STREAK:** Colorless to gray

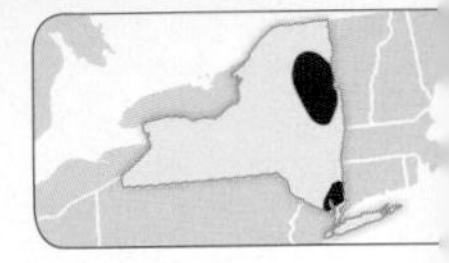
Primary Occurrence

ENVIRONMENT: Mines, outcrops, road cuts, mountains, hills, rivers

WHAT TO LOOK FOR: Hard, yellow-green elongated, striated (grooved) crystals, or brittle masses in rock

SIZE: Crystals tend to be smaller than your thumbnail, but masses can rarely be up to fist-sized

COLOR: Commonly yellow-green; also dark green to brown-green

OCCURRENCE: Common

NOTES: A perennial favorite among collectors, epidote is a fairly common and easily identified mineral found in many places throughout New York State. Unlike most other minerals, epidote's color is a fairly diagnostic trait, as it frequently exhibits a characteristic yellow-green color, sometimes described as "pistachio green," and this is typically your first clue about a specimen's identity. As with most other minerals, fine crystals are rare, but they are occasionally found and take the form of rectangular prisms, typically elongated, with angled, pointed tips and striated (grooved) sides. Sometimes crystals are more needle-like in nature and can be found in radial groupings. But massive specimens or crusts are most common, and they are identified by their hardness, color, indications of crystal structure (most often evident as a fibrous texture), and by the rocks in which they are found. Since it is typically a result of metamorphism, epidote forms in gneiss, schist, and skarn and can be found alongside feldspars, dolomite, among other minerals.

WHERE TO LOOK: Epidote is widespread in small amounts throughout New York. The Tilly Foster Iron Mine near Brewster in Putnam County has produced very fine specimens. Outcrops or rivers in the Adirondacks also yield specimens.

Cluster of fine, blocky albite crysta

Feldspar (orange) in granite

Square cleavage

Well-formed microcline crystal cluster

Feldspars (gray) in anorthosite

Feldspar group

HARDNESS: 6-6.5 **STREAK:** White

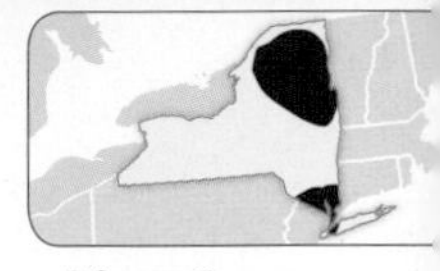

Primary Occurrence

ENVIRONMENT: All environments

WHAT TO LOOK FOR: Abundant, hard, generally light-colored masses in granite; blocky, angular crystals within cavities

SIZE: Varies greatly; typically small but rarely up to a foot or more

COLOR: White to gray, tan to yellow or brown, pinkish to reddish; very rarely blue-green

OCCURRENCE: Very common

NOTES: The feldspar group is a large family of minerals that act as one of the major building blocks of our world; they are present in nearly every type of rock and make up nearly 60 percent of the earth's crust. "Feldspar" is a general name for any member of the group, but the group itself is divided into two subgroups: potassium feldspars, such as microcline, and orthoclase, and plagioclase feldspars, such as albite and anorthite. Generally speaking, potassium feldspars are most abundant in light-colored rocks such as granite, and appear as white to pink, opaque, hard, blocky grains or masses; plagioclase feldspars are most often seen in darker rocks, such as gabbro, as glassier grains. No feldspars are abundant as fine free-standing crystals, but they can still be found and are very blocky with low-angled tips, often alongside amphiboles (page 39). Identifying feldspars isn't too difficult; their hardness is quite high, they're typically rectangular in shape even when embedded, they often exhibit a pearly luster or internal schiller, and they have very square cleavage, which means they break in blocky, nearly 90-degree shapes.

WHERE TO LOOK: Feldspars are common throughout New York; rivers in the Adirondacks yield rocks containing embedded crystals. Free-standing crystals are rarer, but they are found in the Warwick area amid the metamorphic rocks there.

Water-clear fluorite cube from Walworth Quarry

Purple octahedral crystal

Fluorite

HARDNESS: 4 **STREAK:** White

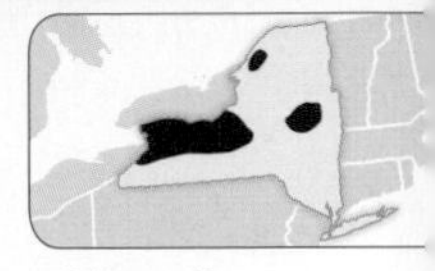
Primary Occurrence

ENVIRONMENT: Lowlands, hills, mines, quarries, outcrops, road cuts

WHAT TO LOOK FOR: Glassy, translucent masses or cubic crystals that can be scratched by a knife but not by a U.S. nickel

SIZE: Crystals are typically smaller than a thumbnail, but masses may be several inches

COLOR: Colorless to white or gray, blue to purple, green to yellow

OCCURRENCE: Uncommon

NOTES: Extremely popular and highly collectible, fluorite is the most common fluorine-bearing mineral and frequently occurs as beautifully formed crystals. Though it forms in a variety of environments, fluorite is most abundant in sedimentary areas where it can often be found within vugs (irregular cavities) in limestone where it typically crystallizes as cubes, often intergrown and clustered together. In other settings, octahedral crystals (which resemble two pyramids placed base-to-base) are more common. Fluorite is always glassy and translucent, but one particular New York locality, the Walworth Quarry, produces fluorite that exemplifies those traits, with specimens that form perfect cubes and with a pristine clarity. More poorly formed crystals or irregular masses may be easy to confuse with other light-colored minerals from sedimentary areas, such as calcite (page 63) and baryte (page 55), but both of those minerals are notably softer. Metallic ore mines also produce fluorite, often in more exciting colors; purple is common.

WHERE TO LOOK: The Walworth Quarry near Rochester has produced water-clear crystals, and fine purple crystals were recovered at quarries near Niagara Falls. But any limestone in Niagara and Monroe Counties could be lucrative.

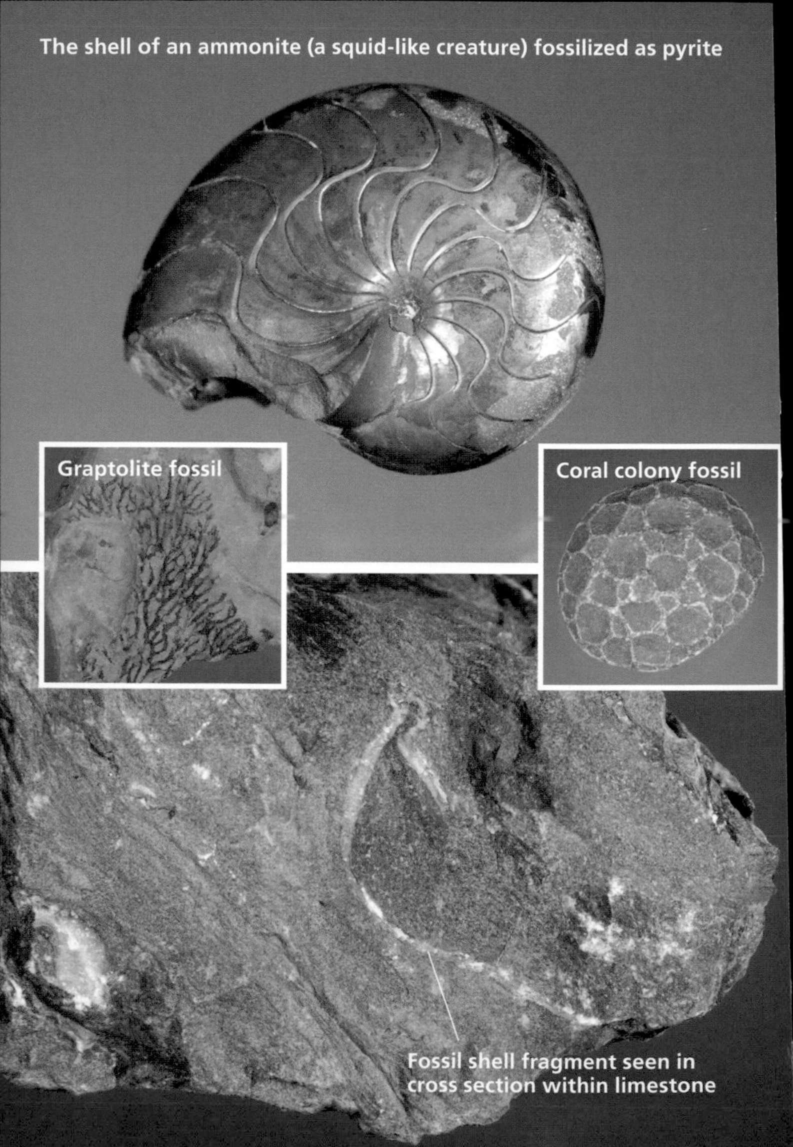

The shell of an ammonite (a squid-like creature) fossilized as pyrite

Graptolite fossil

Coral colony fossil

Fossil shell fragment seen in cross section within limestone

Fossils

HARDNESS: N/A **STREAK:** N/A

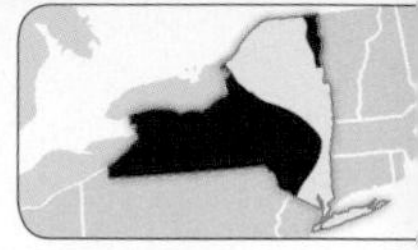

Primary Occurrence

ENVIRONMENT: Lowlands, hills, quarries, rivers, outcrops, road cuts

WHAT TO LOOK FOR: Sedimentary rocks containing embedded mineral structures resembling plants and/or animals

SIZE: Fossils can be any size; size is determined by the organism

COLOR: Tan to brown or gray to black; colored like the surrounding rock

OCCURRENCE: Common to very rare, depending on type

NOTES: New York State is largely overlain by sedimentary rocks, particularly limestone (page 149) and shale (page 201), which formed in ancient seas and lakes that teemed with life millions of years ago. With so much of this potentially fossil-bearing material in the state, New York has become a fossil collectors' destination and has contributed greatly to fossil research. Fossils form when organic matter, such as a shell or leaf, is buried in sediment, particularly underwater, where it cannot decay normally. Over millions of years, minerals from the surrounding rock react with the once-living material and gradually replace its cells, turning it into a mineral formation. This process was more likely to occur with the hard parts of animals, such as snail shells, since they were more apt to survive burial and fossilization. In New York, mostly animal fossils are present, and are most easily found in between the layers of shale. Some are easily identified by appearance alone, while others will require extensive research.

WHERE TO LOOK: Any limestone- or shale-rich region will potentially yield fossils. The Catskills and Finger Lakes areas are rich with fossiliferous shale, as is the far western portion of the state, near Buffalo.

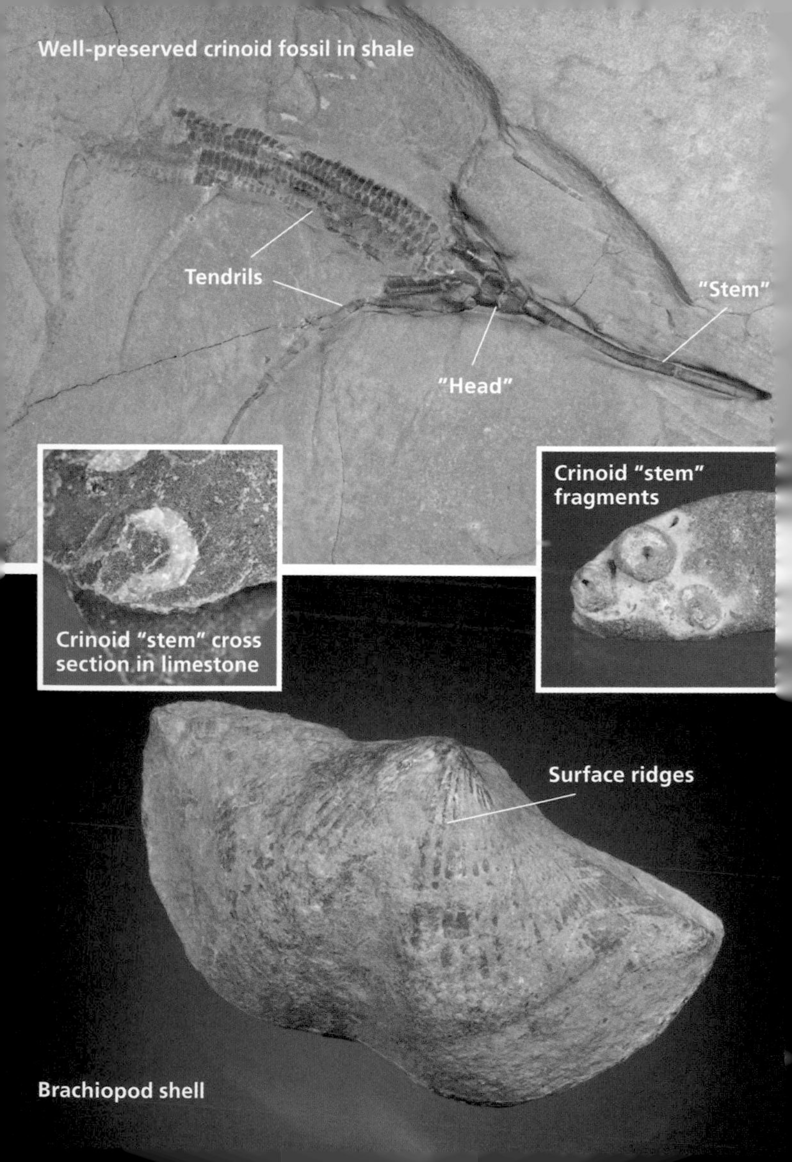
Well-preserved crinoid fossil in shale
Tendrils
"Stem"
"Head"
Crinoid "stem" cross section in limestone
Crinoid "stem" fragments
Surface ridges
Brachiopod shell

Fossils, aquatic animals

HARDNESS: N/A **STREAK:** N/A

Primary Occurrence

ENVIRONMENT: Lowlands, hills, quarries, rivers, outcrops, road cuts

WHAT TO LOOK FOR: Sedimentary rocks containing traces of shells or tubular structures

SIZE: Aquatic animal fossils can reach a foot long and rarely larger

COLOR: Tan to brown or gray to black; colored like the surrounding rock

OCCURRENCE: Uncommon

NOTES: Because of the repeated submersion of ancient New York beneath seas and oceans, a seemingly endless variety of aquatic animal fossils are present in the state's sedimentary rocks. Individual animal fossils are less abundant than those of colonial animals (discussed on page 101), but they are still widespread in New York, and include animals like crinoids, bivalves, and brachiopods. All generally were best-preserved in the state's shale (page 201) formations and many are fairly conspicuous and make for exciting finds. Crinoids, nicknamed "sea lilies," are a class of animals that still exist today; they anchor themselves with a stalk while their "head" branches out with fan-like structures to catch food; as fossils, round, tubular disjointed segments of crinoid stalks are most common, often appearing as odd disk-shaped objects embedded in rock, but fully intact specimens have been found in New York. Bivalves and brachiopods are easily spotted and identified as well, due to their striking resemblance to modern-day clams and other shellfish.

WHERE TO LOOK: Any limestone- or shale-rich region could yield fossils. The Catskills and Finger Lakes areas are rich with fossiliferous shale, as is the far western portion of the state, near Buffalo.

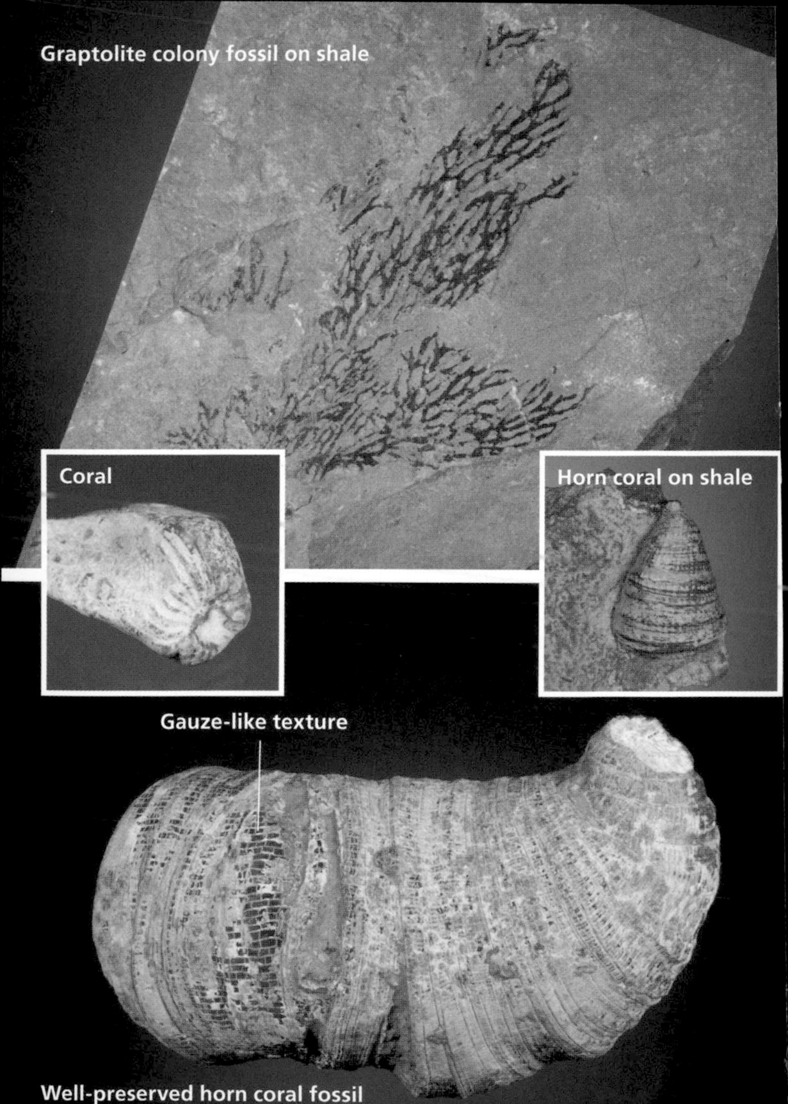
Graptolite colony fossil on shale
Coral
Horn coral on shale
Gauze-like texture
Well-preserved horn coral fossil

Fossils, colonial animals

HARDNESS: N/A **STREAK:** N/A

Primary Occurrence

ENVIRONMENT: Lowlands, hills, quarries, rivers, outcrops, road cuts

WHAT TO LOOK FOR: Sedimentary rocks containing traces of conical or branching structures

SIZE: Colonial animal fossil specimens tend to be up to palm-sized and rarely larger

COLOR: Tan to brown or gray to black; colored like the surrounding rock

OCCURRENCE: Common to uncommon, depending on species

NOTES: Though many of the aquatic fossils found in New York consist of individual animals (page 99), many more are the remains of colonial animals like coral. Colonial animals are often nearly microscopic and live together in large numbers; the resulting colonies take shape as a structure of some sort. Coral, particularly horn coral, is an especially abundant example in New York, and appears as tubular, curving structures with a gauze- or wood-like texture. As their name suggests, these fossils are actually horn-shaped, and they are conspicuous when found embedded in shale or even when found loose after being weathered free. Other, less common corals may be more branching in shape, perhaps initially similar to crinoids (page 99), but they can be differentiated by their non-segmented structure and gauzy texture. Graptolites are seemingly plant-like animal colonies that anchored on the seafloor and branched outward; fossils of this extinct group of animals are uncommon and appear as very thin impressions in between layers of shale.

WHERE TO LOOK: Much of the western portion of the state, particularly in the Buffalo area, is known for fossil-bearing shales that are particularly rich with horn corals.

Very well-preserved *eurypterus remipes* in shale

Paddle-like leg

Tail spine

Eyes

Body fragment

Walking legs

Swimming legs

Eurypterus remipes (missing tail)

Fossils, eurypterids

HARDNESS: N/A **STREAK:** N/A

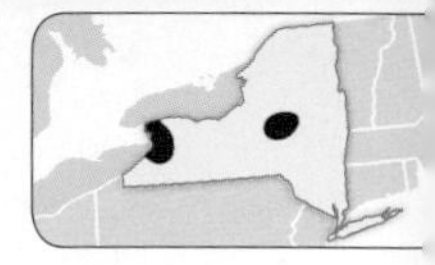
Primary Occurrence

ENVIRONMENT: Lowlands, hills, quarries, rivers, outcrops, road cuts

WHAT TO LOOK FOR: Sedimentary rocks containing traces of unusual crab-like creatures with pointed tails

SIZE: Specimens can be up to a foot in size and rarely larger

COLOR: Tan to brown or gray to black

OCCURRENCE: Very rare

NOTES: The arthropods are a group of animals that includes crustaceans, insects, and arachnids as well as some organisms that are now long extinct such as eurypterids. Nicknamed "sea scorpions" due to their menacing appearance, the eurypterids were predators of the ancient seas (and later, of freshwater), and were equipped with hard-shelled, segmented bodies, a spear-like tail, sets of legs for walking on the seafloor, and a large paddle-shaped pair of legs for swimming. Though somewhat crab-like in nature, their tapered, teardrop-shaped bodies are distinctive and distinguish them from trilobites (page 109), even when their large swimming legs or tail spike are missing (as they often are). They were first discovered in New York, and one specific species, *Eurypterus remipes*, has made such a scientific impact in the state that it has been named the New York state fossil. But eurypterids all went extinct long ago during an extinction event at the end of the Permian geological period.

WHERE TO LOOK: The western portion of New York, particularly Erie and Niagara Counties, has been the site of the most significant finds, including the first fossil discoveries of the animals. Look in Herkimer County and near the many the Buffalo area quarries and shale exposures.

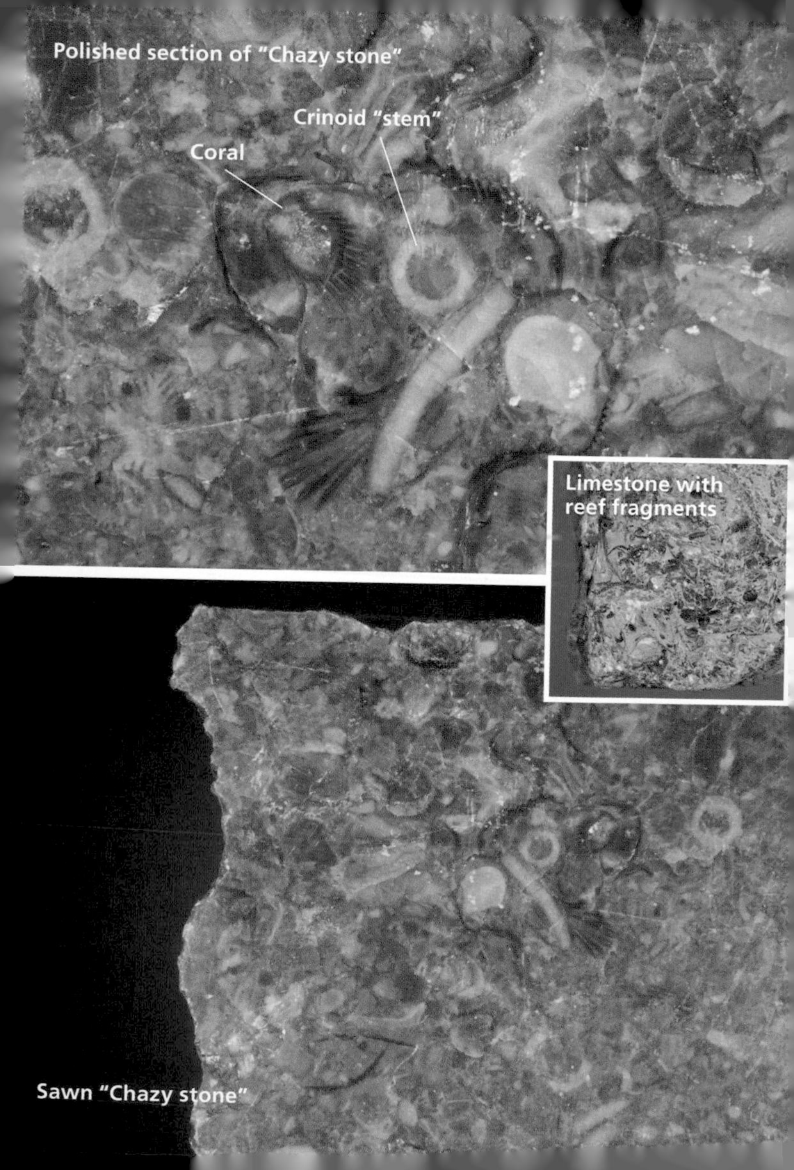

Polished section of "Chazy stone"
Crinoid "stem"
Coral
Limestone with reef fragments
Sawn "Chazy stone"

Fossils, reef

HARDNESS: N/A **STREAK:** N/A

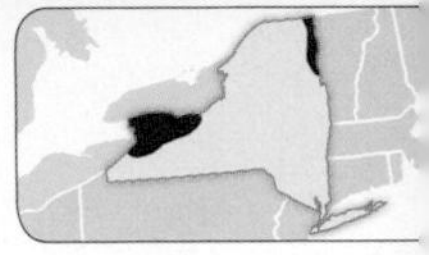
Primary Occurrence

ENVIRONMENT: Lowlands, hills, quarries, outcrops, rivers

WHAT TO LOOK FOR: Limestone containing numerous embedded fossils of varying organisms

SIZE: Fossil reef material is a rock that can be found in any size

COLOR: Varies greatly; multicolored, predominantly gray to black, with white, pink, and brown spots

OCCURRENCE: Rare

NOTES: The Ordovician geological period began 485 million years ago. Then, as now, the ancient earth's oceans teemed with life and New York was situated beneath a warm sea and was home to muddy reefs swarming with creatures. When hard-shelled animals like snails died, their rigid fragments, which consisted largely of the mineral aragonite (page 49), often settled into large beds and intermixed with the remains of colonial animals such as coral. These beds eventually solidified and compacted, locking together with tiny crystallized calcite (page 63) grains derived from the aragonite shells. The result is limestone (page 149) that is abounding with fossil fragments to the extent that a specimen looks like a mosaic of ancient life. This material can be very attractive as a collectible, especially when the fossils are colorful and well preserved as in "Chazy stone," a popular variety found in the state. Like other limestones, fossil reef material is soft and easily dissolves in undiluted vinegar, with the only difference being the concentration of fossil material.

WHERE TO LOOK: The Champlain Valley in northeastern New York, particularly near West Chazy, is where the Chazy Reef Formation is located, for which "Chazy stone" is named.

Worm tracks in shale

Worm track in shale

Trace fossil

Worm burrows in shale

Fossils, trace

HARDNESS: N/A **STREAK:** N/A

Primary Occurrence

ENVIRONMENT: Lowlands, hills, quarries, rivers, outcrops, road cuts

WHAT TO LOOK FOR: Sedimentary rocks containing tracks, tunnels, or paths created by ancient life

SIZE: Trace fossils can be any size; most are under an inch or two

COLOR: Tan to brown or gray to black; colored like the surrounding rock

OCCURRENCE: Uncommon to very rare, depending on type

NOTES: For most people, the word fossil refers to a piece of an ancient organism, such as a bone, a shell, or a leaf, embedded in rock. But some fossils are far more subtle, giving us a glimpse not of an ancient creature itself but of its behavior. These are called trace fossils, and as their name implies, they exhibit a preserved trace of an animal rather than its body parts. These preserved traces include footprints, tracks, burrows, and even feces. Several kinds of trace fossils are present in New York, but few are abundant except for the burrows and tracks of small aquatic animals such as mollusks and worms. Most of the time, these tunnel-like burrows are not hollow, but are slight depressions or raised vein-like structures in the rock and formed when slightly harder sediment filled in the holes. You'll most easily find these in the layers of shale (page 201) but also in limestone (page 149). These fossils are tricky to identify, so research and expert help may be needed.

WHERE TO LOOK: The Finger Lakes region, particularly Cayuga Lake near Ithaca, is lined with shales containing various tracks and burrow holes, which are easily observed at many rock exposures. Far rarer mollusk tracks resembling tire treads have been found in Clinton County.

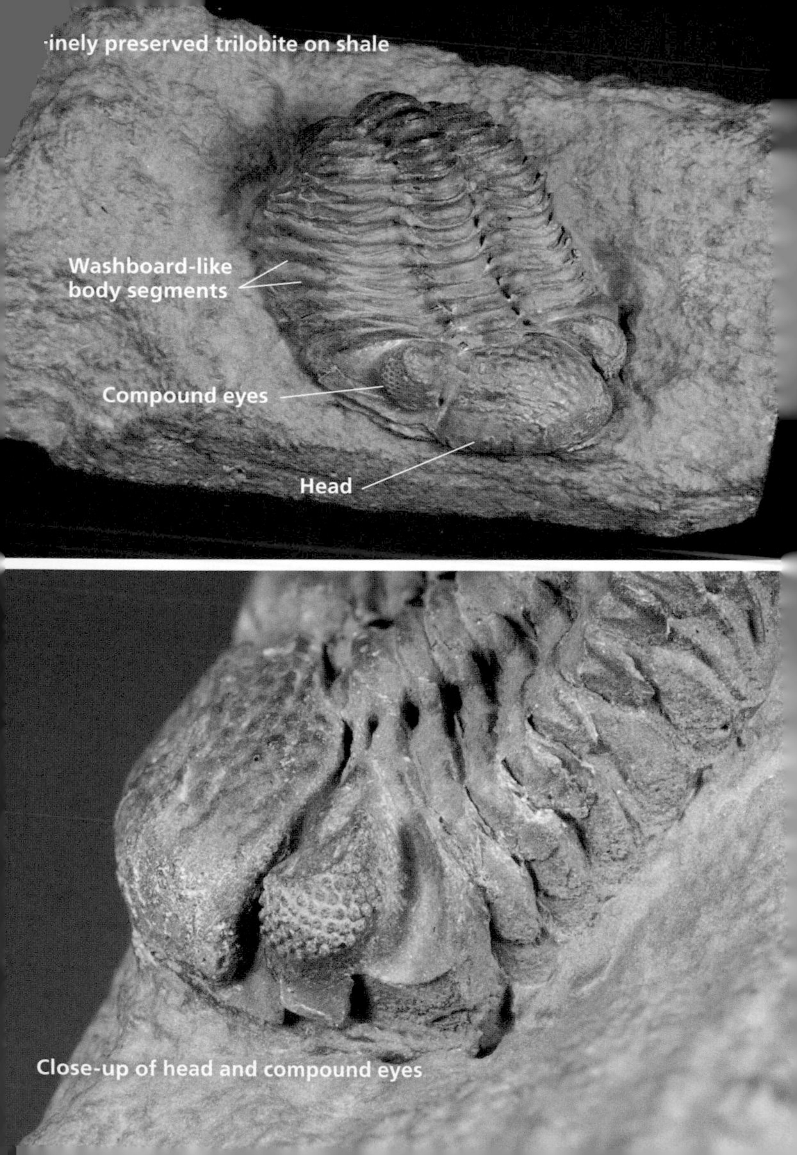

-inely preserved trilobite on shale

Close-up of head and compound eyes

Fossils, trilobites

HARDNESS: N/A **STREAK:** N/A

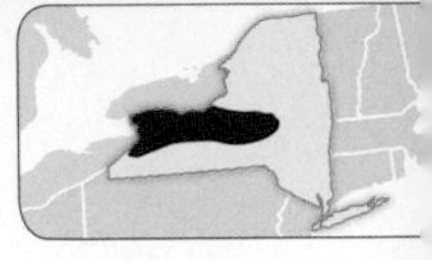
Primary Occurrence

ENVIRONMENT: Lowlands, hills, quarries, rivers, outcrops, road cuts

WHAT TO LOOK FOR: Sedimentary rock containing traces of segmented, almost insect-like animals with bulbous eyes and spikes

SIZE: Most trilobites are smaller than your palm

COLOR: Tan to brown or gray to black

OCCURRENCE: Rare

NOTES: Though many kinds of fossils are found throughout New York, few are as striking as a well-preserved trilobite. These creatures were present in earth's seas and oceans for nearly 300 million years, but this class of arthropod was unable to survive a mass-extinction event at the end of the Permian geological period, over 250 million years ago. Their closest living relatives today are crustaceans and arachnids. Trilobites had hard-shelled bodies that consisted of several armor-like segments and a head with two bulbous compound eyes. As they evolved, their bodies became more complex, developing features like defensive spikes. As fossils, they are often found in the layers of shale (page 201) as well as embedded in limestone (page 149). They are conspicuous when found whole, with oval-shaped bodies consisting of repeating ridge-like segments and a horse-shoe- or dome-shaped head with two eye bumps. But many partial specimens can be found as well, and they are more difficult to identify. Compare with eurypterids (page 103).

WHERE TO LOOK: The beds of shale from central New York, around Utica, west to the Niagara area are prolific for fossils of all kinds, including well-preserved trilobites. Quarries in the Lockport area in Niagara County are well known.

Gabbro
Texture detail
River-worn specimen
Rough gabbro

Gabbro

HARDNESS: >5.5 **STREAK:** N/A

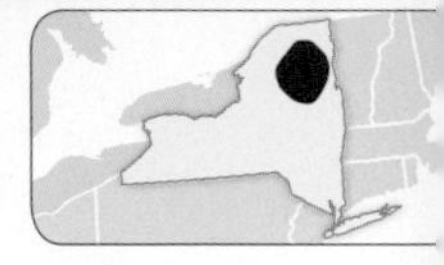

Primary Occurrence

ENVIRONMENT: Mountains, hills, outcrops, road cuts, rivers

WHAT TO LOOK FOR: Dark, very coarse-grained, dense rocks with visible glassy, often rectangular crystals

SIZE: As a rock, gabbro can be found in any size

COLOR: Mottled but predominantly dark gray to black with a greenish tint, with lighter spots; browner when weathered

OCCURRENCE: Uncommon

NOTES: Portions of the Adirondack Mountains are made up of gabbro, which is an igneous rock that is virtually identical to basalt and diabase (page 57) in composition, but gabbro formed deep within the earth, rather than at or near the earth's surface. This means that as the magma (molten rock) from which the gabbro formed cooled, it did so very slowly, allowing the minerals within it to crystallize to a large, easily visible size. Consisting primarily of plagioclase feldspar, olivine, pyroxenes, and magnetite (mostly dark-colored minerals), gabbro is always nearly black in appearance, often with a greenish tint in bright light. Upon close observation you can often easily find well-formed, rectangular feldspar crystals among its coarse grains, which are typically more reflective than the rest of the rock. You won't confuse it with its fine-grained analogs, basalt and diabase, as they don't occur in the same regions, but you could confuse gabbro with anorthosite (page 43). Anorthosite contains fewer dark minerals, thereby appearing much lighter in color.

WHERE TO LOOK: The Adirondacks are your best bet for gabbro; rivers and outcrops in the Lake Placid and Whiteface Mountain area yield collectible chunks. Other northern parts of the state may yield pieces in gravel deposited by glaciers.

Brightly lustrous, freshly broken galena

Tiny galena crystal

Smooth galena mass

Cubic galena crystal cluster with sphalerite (brown)

Galena

HARDNESS: 2.5 **STREAK:** Lead gray

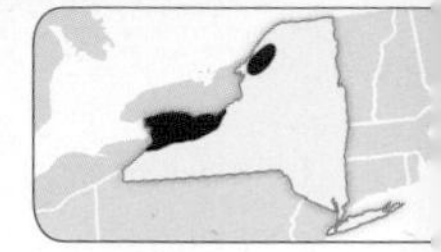
Primary Occurrence

ENVIRONMENT: Mines, quarries, outcrops, road cuts, hills, lowlands

WHAT TO LOOK FOR: Soft but very heavy dark, metallic mineral, often very blocky in shape and found in sedimentary rocks

SIZE: Individual crystals are up to fist-sized, usually under an inch

COLOR: Dark lead-gray, often with white to light gray or tan-brown surface coating; bright silvery blue when broken

OCCURRENCE: Common

NOTES: Galena consists of lead and sulfur, and though you'll want to wash your hands afterwards, it is still largely safe to handle despite its lead content, so don't let that deter you. It is the world's primary ore of lead and was mined in many areas throughout New York State, generally occurring alongside sphalerite (page 213), chalcopyrite (page 75), and other sulfur-bearing minerals. Often found in sedimentary rocks like limestone or in ore veins deposited in other types of rock by rising warm water, galena crystals are conspicuous: typically perfectly cubic in shape, often intergrown in clusters, and dull gray in color. In addition, it has cubic cleavage, which means that when broken, blocky or step-like surfaces will appear, and when freshly broken it appears very brightly lustrous with a metallic silver-blue color. Older surfaces darken in color, often developing white or tan crusts of cerussite (page 69). Galena also has a very high specific gravity, which means that it is so dense that even a small specimen feels heavy for its size.

WHERE TO LOOK: Some of the finest specimens have come from the Balmat-Edwards and Rossie areas in St. Lawrence County. The sedimentary rock quarries of Niagara and Monroe Counties are also known for small but perfect crystals.

Gore Mountain garnet (red) in amphibolite
Almandine crystal
Garnet on schist
Garnets (red) in granite

Garnet group

HARDNESS: 6.5–7.5 **STREAK:** Colorless

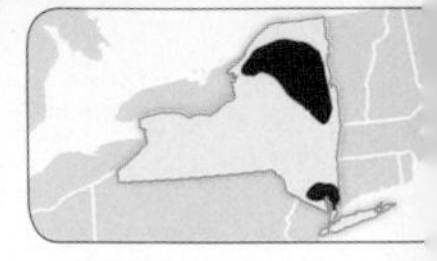

Primary Occurrence

ENVIRONMENT: Mountains, hills, outcrops, road cuts, mines, rivers

WHAT TO LOOK FOR: Very hard, small, ball-like crystals embedded in granite or schist

SIZE: Garnets are typically smaller than an inch

COLOR: Red to dark red or brown, green, rarely black

OCCURRENCE: Common

NOTES: Consisting of several minerals of similar composition and structure, the garnet group is a large family of minerals that form in a number of ways, but most often occur as a constituent of coarse-grained igneous rocks like granite or as a product of metamorphism in rocks like schist, gneiss, and skarn. Mostly reddish brown in color, garnets generally form as rounded ball-like crystals that exhibit many facets when finely developed. New York is home to several species of garnet, including andradite, grossular, and spessartine, but almandine is by far the most prominent, particularly throughout the Adirondack Mountains. Telling them apart can be difficult and only expert collectors can identify them, but differentiating a garnet from other minerals is easy. Their ball-like shape is distinctive, as is their hardness and color; similarly colored vesuvianite (page 237) and titanite (page 225) may, if poorly formed, be confused with garnets, but neither are as hard. Lastly, garnets are so hard that they often able to survive weathering that destroys their host rock, so they are frequently found loose in river gravel.

WHERE TO LOOK: New York's state gem is the Gore Mountain garnet from near North River, where it can be accessed via a pay-to-dig service, but the entire Adirondack region is prolific with garnets. Riverbeds yield tiny loose garnets.

River-worn gneiss with tiny garnets (red)

River-worn gneiss with tiny garnets (red) and pyroxenes (black)

Gneiss

HARDNESS: N/A **STREAK:** N/A

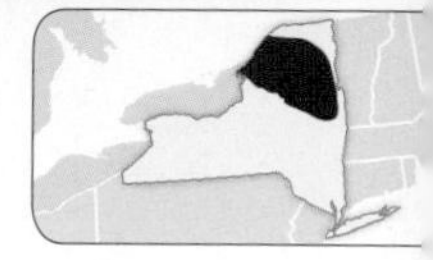

Primary Occurrence

ENVIRONMENT: All environments

WHAT TO LOOK FOR: Loosely layered, streaked, or banded rocks, often with lots of quartz, micas and garnets

SIZE: As rocks, gneisses can be found in any size

COLOR: Varies greatly; typically multicolored in white to grays or black, greens to browns, and reds

OCCURRENCE: Common

NOTES: When rocks like granite are subjected to heat and/or pressure after their initial formation, they undergo physical and chemical alterations in a process called metamorphism. During metamorphism, the rock is often partially melted or otherwise made semi-fluid, causing deformation of its original structure as well as causing the minerals within it to change and recombine, often into completely new ones. Eventually, when the rock resolidifies, the resulting rock is new. It may still somewhat resemble the original type, but with key differences such as layering or the clustering of similar mineral grains. Gneiss (pronounced "nice") is the quintessential example; it is a general term used to describe rocks with loosely defined layers of minerals that were rearranged and compressed. Traditionally described as having less than half of its minerals concentrated into layers, gneiss retains much of the original rock's appearance, often with the inclusion of new minerals such as garnet. Varieties are even named for their parent rock, so gneiss resulting from granite, for example, is called granitic gneiss. Compare and contrast gneiss with schist on page 195.

WHERE TO LOOK: Gneiss is abundant in the Adirondack Mountains where specimens can be found in nearly any river. It is also common in gravel, due to the glaciers that deposited it.

Various river-worn granites and granitoids
Texture detail
Texture detail
Various granites and granitoids

Granite

HARDNESS: N/A **STREAK:** N/A

Primary Occurrence

ENVIRONMENT: Mountains, hills, mines, quarries, road cuts, outcrops, rivers

WHAT TO LOOK FOR: Coarse-grained rock containing grains of many different minerals, each visible with the naked eye

SIZE: As a rock, granite can occur in any size

COLOR: Varies greatly; multicolored, primarily in shades of white to tan or pink, gray to black, brown, red to orange, green

OCCURRENCE: Common

NOTES: The foremost example of an igneous rock, granite forms when magma (molten rock) buried deep within the earth cools very slowly. This allows the minerals contained within it ample time to crystallize to a large, visible size, which gives granite its characteristic mottled appearance. Each mineral in granite—primarily quartz, orthoclase feldspar, amphiboles, and micas—can be seen embedded in the rock as grains of more-or-less equal size, with some, particularly the feldspars and micas, frequently clearly exhibiting their crystal shapes. Granite is predominantly light-colored with lots of quartz and feldspars and fewer dark minerals, but as that ratio changes, it is then classified as a different type of rock. These granite-like rocks are called granitoids, and due to their similarities to granite and the difficulty of distinguishing them, they typically all get lumped under the "granite" name. In New York, granite could be confused with anorthosite (page 43), which is lighter in color, or gabbro (page 111), which is darker.

WHERE TO LOOK: Granites are abundant throughout the Adirondacks, found along any river as worn pebbles. Large amounts of granite were also deposited all over the state by the glaciers, so most rivers or gravel will yield specimens.

Graphite (silvery gray) on marble

Fine ¼" crystal

Fine ¼" crystal

Tiny silvery gray graphite flakes in metamorphic rock

Graphite

HARDNESS: 1-2 **STREAK:** Black to steel gray

Primary Occurrence

ENVIRONMENT: Mines, quarries, road cuts, hills, mountains, outcrops

WHAT TO LOOK FOR: Very soft, flexible, hexagonal (six-sided), flat crystals, typically embedded in marble

SIZE: Crystals are usually less than ¼ inch, up to thumbnail size

COLOR: Metallic black to steel gray, sometimes with a bluish tinge

OCCURRENCE: Uncommon

NOTES: Like gold or silver, graphite is an example of a native element, or a mineral consisting of a single pure element, and in this case, graphite consists of just carbon. Carbon is one of a handful of elements with multiple allotropes, or structural variations of a crystallized element; for example, another native form of pure carbon is diamond. But where diamond is the hardest known naturally occurring mineral, graphite is one of the softest; this discrepancy is due to their differing crystalline structural forms. Graphite develops most often as brightly lustrous, opaque, metallic hexagonal (six-sided) plate-like crystals that are very thin and exhibit a stacked, layered nature, often in marble (page 157). These crystals are very flexible and so soft that their layers can be peeled apart. They'll easily leave a streak, even on paper (hence graphite's use as pencil lead). These traits are fairly distinctive, but graphite still resembles micas (page 161), though micas are not metallic and are more common, and molybdenite, a rarer New York mineral that will streak gray-green on a streak plate but black on paper.

WHERE TO LOOK: There are numerous graphite localities in the state, the best of which are in Orange County, particularly around the town of Warwick, and in Essex County, at old mine sites near Newcomb and Ticonderoga.

Fine ⅛" groutite crystals (black) on talc

Crystals in pocket

1⁄16" groutite crystals (black) in talc

Groutite

HARDNESS: 3.5-4 **STREAK:** Dark brown

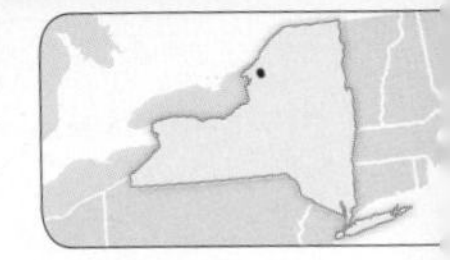
Primary Occurrence

ENVIRONMENT: Mines

WHAT TO LOOK FOR: Tiny, metallic, black needles within cavities in talc from St. Lawrence County

SIZE: Crystals are very small, typically shorter than ¼ inch

COLOR: Metallic black

OCCURRENCE: Very rare

NOTES: A manganese hydroxide, groutite is one of New York's rarest minerals, but it is also one of the better known to collectors. That's because of its unusual occurrence in the state; it is found within a talc-rich schist, which is fairly unique for the mineral throughout the world, and when found it typically has a finely crystallized appearance. Closely related to goethite (see limonite, page 151), groutite usually forms lens-shaped blade-like crystals, but not in New York; groutite from the state appears as tiny, delicate metallic needles with angled tips and striated (grooved) sides, often grown in small clusters. But even without testing the little crystals for hardness or streak, you'll be able to identify them because groutite is only found at a single locality in New York and nothing else from the area will resemble it. Crystals will be found in small pockets and cavities within the talc (page 221) from the area, alongside few other metallic minerals, though pyrite (page 177) and magnetite (page 155) are rarely found there, too. Pyrite forms brassy, cubic crystals, though, and magnetite is magnetic while groutite is not.

WHERE TO LOOK: One single New York locality produced this rare mineral, the "No. 2½ Mine" in Talcville, St. Lawrence County. Though the mine is largely gone and off-limits to collectors who haven't obtained permission, specimens are popular and are generally always available on the market.

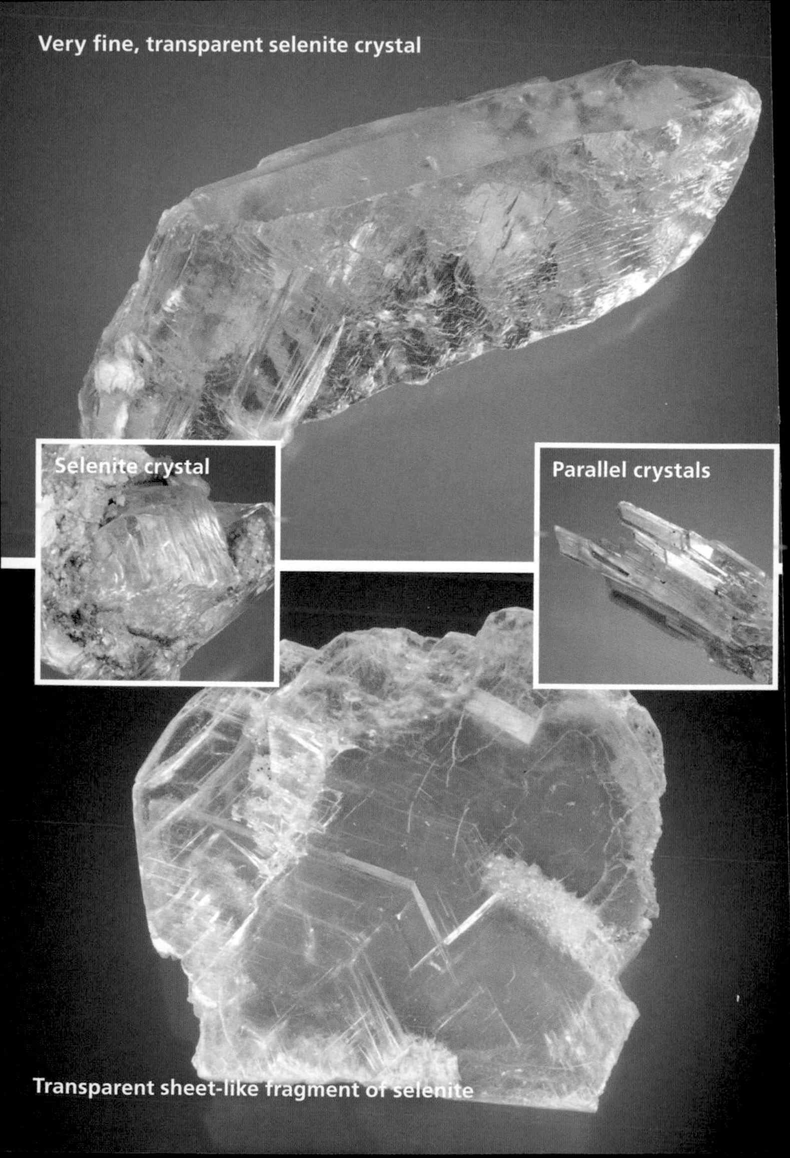
Very fine, transparent selenite crystal

Selenite crystal

Parallel crystals

Transparent sheet-like fragment of selenite

Gypsum

HARDNESS: 2 **STREAK:** White

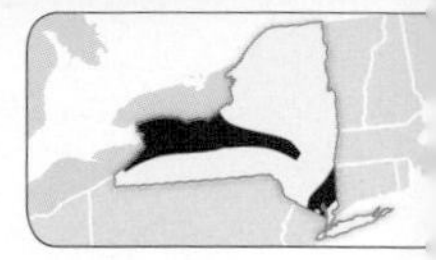

Primary Occurrence

ENVIRONMENT: Mines, quarries, lowlands, hills, rivers, road cuts, outcrops

WHAT TO LOOK FOR: Very soft, light-colored grainy masses or glassy angular crystals easily scratched by a fingernail

SIZE: Crystals tend to be smaller than your palm, but masses can be several feet in size

COLOR: Colorless to white or gray; yellow to brown when impure

OCCURRENCE: Common

NOTES: The world's most common sulfur-bearing mineral and the primary ingredient in plaster for thousands of years, gypsum can form in multiple ways, in many environments, and can take on numerous appearances. Most common within limestone in sedimentary regions where long-gone seas dried up and left behind minerals, gypsum can be found both massively as enormous beds between rock formations or as fine crystals grown in cavities. Massive varieties tend to be granular in texture, chalky to the touch, and opaque, but selenite, the name given to well-crystallized varieties, is glassy, translucent, and is found as delicate needles, window-like sheets, or angular prisms. No matter its appearance, gypsum is always very soft and easily scratched with just a fingernail. Gypsum often resembles calcite (page 63), dolomite (page 87), and aragonite (page 49), with which it can occur, but those minerals are considerably harder and are not gradually soluble in water like gypsum. It also can resemble halite (page 127), but halite is harder and dissolves more rapidly.

WHERE TO LOOK: Gypsum is a common find in Niagara, Wayne, and especially Monroe Counties. Particularly fine crystals are collected from quarries near Rochester.

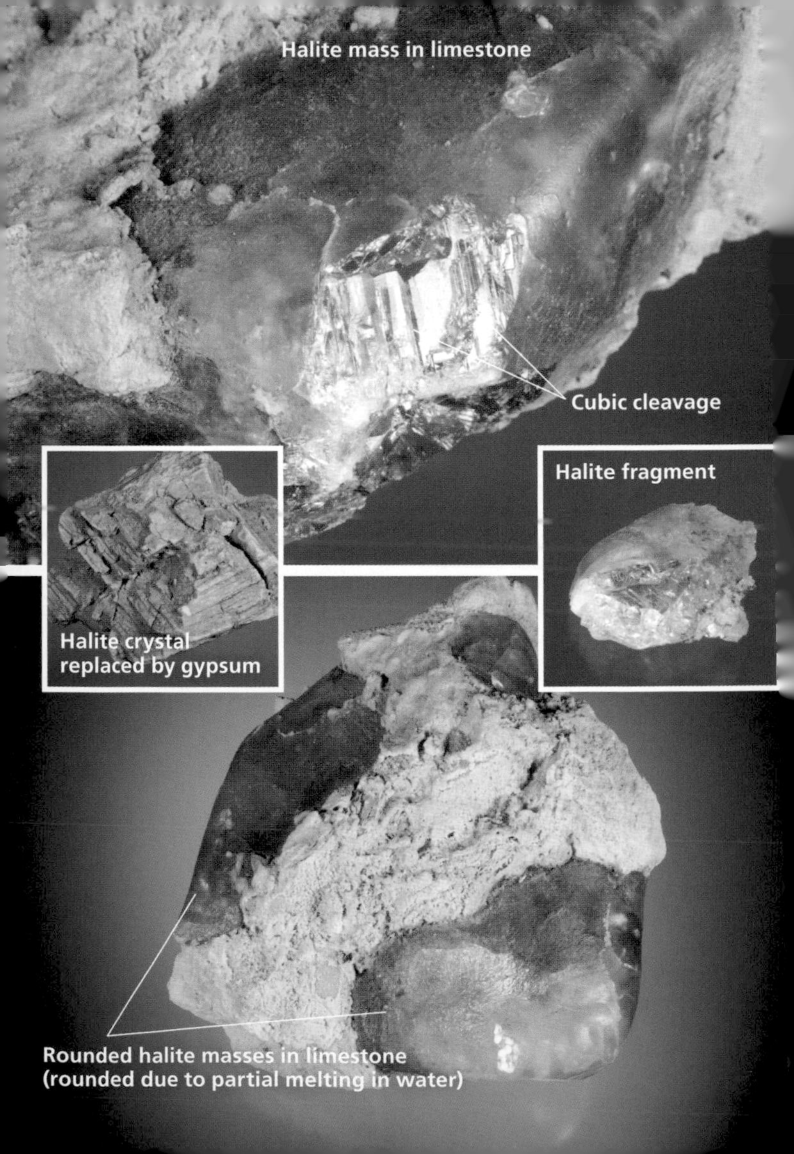

Halite mass in limestone
Cubic cleavage
Halite fragment
Halite crystal
replaced by gypsum
Rounded halite masses in limestone
(rounded due to partial melting in water)

Halite

HARDNESS: 2-2.5 **STREAK:** White

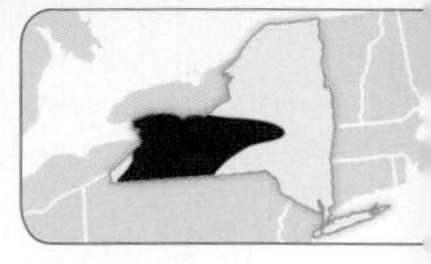

Primary Occurrence

ENVIRONMENT: Lowlands, quarries, mines, rivers

WHAT TO LOOK FOR: Transparent, very soft, cubic crystals or masses that rapidly dissolve in water

SIZE: Halite is found in masses up to several feet in size

COLOR: Colorless to white, gray

OCCURRENCE: Uncommon

NOTES: Many people do not realize that the table salt they use on their food must be mined from the earth, and that "salt" is actually a mineral called halite. Consisting of sodium and chlorine, halite was deposited when ancient oceans dried up; as seawater evaporates, it leaves behind dissolved minerals called evaporites. Halite is the most prominent evaporite worldwide, along with gypsum (page 125), and was often deposited in enormous, thick beds that underlie other sedimentary rocks. As such, crystals are rarely found, especially in New York, and massive or granular chunks are the most common specimens, often adhering to limestone or other sedimentary material. When crystals are present, they appear as glassy perfect cubes, often elaborately intergrown, but crystallized or not, all specimens have cubic cleavage, or will break into cubes when struck. In addition, all halite dissolves rapidly in water, even in the rain, which when partially wetted will give a specimen a "melted" look. Halite can greatly resemble gypsum but is harder.

WHERE TO LOOK: Halite underlies large portions of New York State, particularly in the southern areas, but very little is actually accessible on the surface. The Erie-Ontario Lowlands, which run from Syracuse westward (and includes I-90), may sometimes yield specimens in fresh diggings, especially closer to the Niagara area.

Cluster of finely bladed hematite crystals from Chub Lake
Metallic hematite crust
Hematite mass
Metallic gray hematite
Red weathered surfaces

Hematite

HARDNESS: 5-6 **STREAK:** Reddish brown

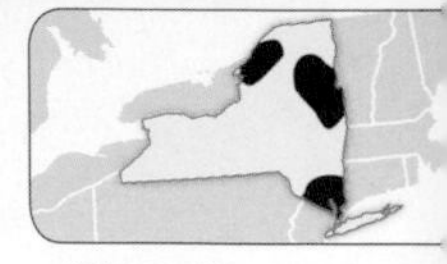
Primary Occurrence

ENVIRONMENT: All environments

WHAT TO LOOK FOR: Dark gray metallic blade-like crystals or masses; also crusts or masses of rust-red material

SIZE: Crystals tend to remain smaller than a thumbnail, but masses can be up to a foot or more

COLOR: Steel-gray to black; rust-red to brown when weathered

OCCURRENCE: Very common

NOTES: Hematite is a simple combination of iron and oxygen and is therefore the most abundant ore of iron on earth, able to form in a wide range of environments and rock types. Crystals are rare, taking the form of thin hexagonal (six-sided) plates, often intergrown and arranged in more-or-less parallel groups, while masses and crusts, often with botryoidal (lumpy, grape-like) surfaces are more abundant. In all cases, specimens are opaque and metallic black unless weathered, in which case rusty red surface coatings will form, both on the hematite and the surrounding rock. Being able to identify hematite is important for new collectors, but it's also fairly easy. Hematite's hardness is distinctive, but its streak color is even more so, appearing reddish no matter how black the specimen may be. This distinguishes it from limonite and goethite (page 151), which have a yellow streak. Magnetite (page 155) and ilmenite (page 141) may also look similar, but both are magnetic while hematite is not.

WHERE TO LOOK: Hematite is ubiquitous and can be assumed present in almost any rock with a reddish color. The Chub Lake area in St. Lawrence County has produced some of the finest crystals in the state, but virtually any outcrop in the state has the potential to produce some amount of hematite.

Oolitic hematite

Close-up of oolitic hematite showing texture of countless spheres

Hematite, varieties

HARDNESS: 5-6 **STREAK:** Reddish brown

Primary Occurrence

ENVIRONMENT: Mines, outcrops, road cuts, hills, mountains

WHAT TO LOOK FOR: Material with the rust-reddish color of weathered hematite; often found as coarse, massive chunks

SIZE: Varieties of hematite can be found in a wide range of sizes, even up to several feet

COLOR: Reddish brown, rust-red

OCCURRENCE: Common to rare, depending on variety

NOTES: Hematite forms so easily and in such a variety of environments that each setting can affect its development or provide a different source of the iron. Limonite (page 151) is a good example; it is a mixture of iron-bearing minerals, including hematite, and is deposited as crusts or coatings on other rocks and minerals as older iron-bearing minerals weather away. Ochre, a chalky red limonite-like material that contains mostly hematite, is a similar case. A more interesting example, however, is oolitic hematite, also called sedimentary hematite, specimens of which are well known from New York. It consists of reddish masses replete with countless oolites, tiny spherical formations. This peculiar variety of hematite formed when limestone was slowly replaced by iron ores, with the oolites being spheres of hematite that nucleated (formed around a central point) around tiny fossils. A hardness test may not apply for some varieties of hematite, but its characteristic rust-red coloration is often the only observation needed.

WHERE TO LOOK: The area around Clinton in central New York is the source of the interesting sedimentary oolitic variety of hematite, while ochre can be found in many areas rich in sedimentary rocks or clay, particularly around lakes.

Very large crystal (approx. 4")
Dolomite crystals
Crystals in vug
Assortment of loose "Herkimer diamonds"

"Herkimer diamonds"

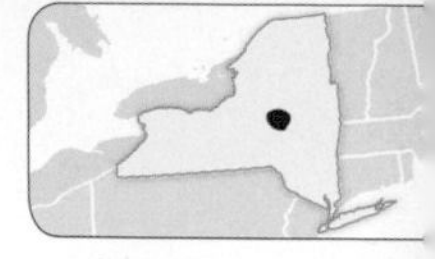
Primary Occurrence

HARDNESS: 7 **STREAK:** White

ENVIRONMENT: Quarries

WHAT TO LOOK FOR: Extremely well-formed and unusually transparent quartz crystals within pockets in limestone in central New York

SIZE: Crystals are rarely larger than an inch; very rarely fist-sized

COLOR: Colorless to white; rarely gray to black

OCCURRENCE: Uncommon

NOTES: Arguably New York's "true" state gem, Herkimer diamonds are an incredibly popular collectible that is fairly abundant—if you're in the right area. Of course, they're not diamonds at all; they're actually quartz (page 185) but of such a remarkable clarity that the "diamond" nickname is apt. Known to American Indians long before the first Europeans arrived in New York, the "herks," as locals know them, are found in pockets within dolostone in the Herkimer area. These crystals developed with double terminations, which means that not one but both ends of a crystal terminate, or have a natural pointed tip. They are often found alongside dolomite (page 87), calcite (page 63), and pyrobitumen (page 137), which is sometimes even trapped within the quartz. How they developed is still somewhat of a mystery; from their clarity, we know they crystallized very slowly, and geologists think they may have formed thanks to an interplay between silica-bearing acids and bacteria. The most prized varieties consist of smoky quartz (gray to black in color) or scepters (a thin crystal topped by a thicker portion).

WHERE TO LOOK: All currently known localities are near Herkimer, Middleville, St. Johnsville, and Fonda in Herkimer and Montgomery Counties. Most localities are part of pay-to-dig businesses, but this ensures that you'll find specimens.

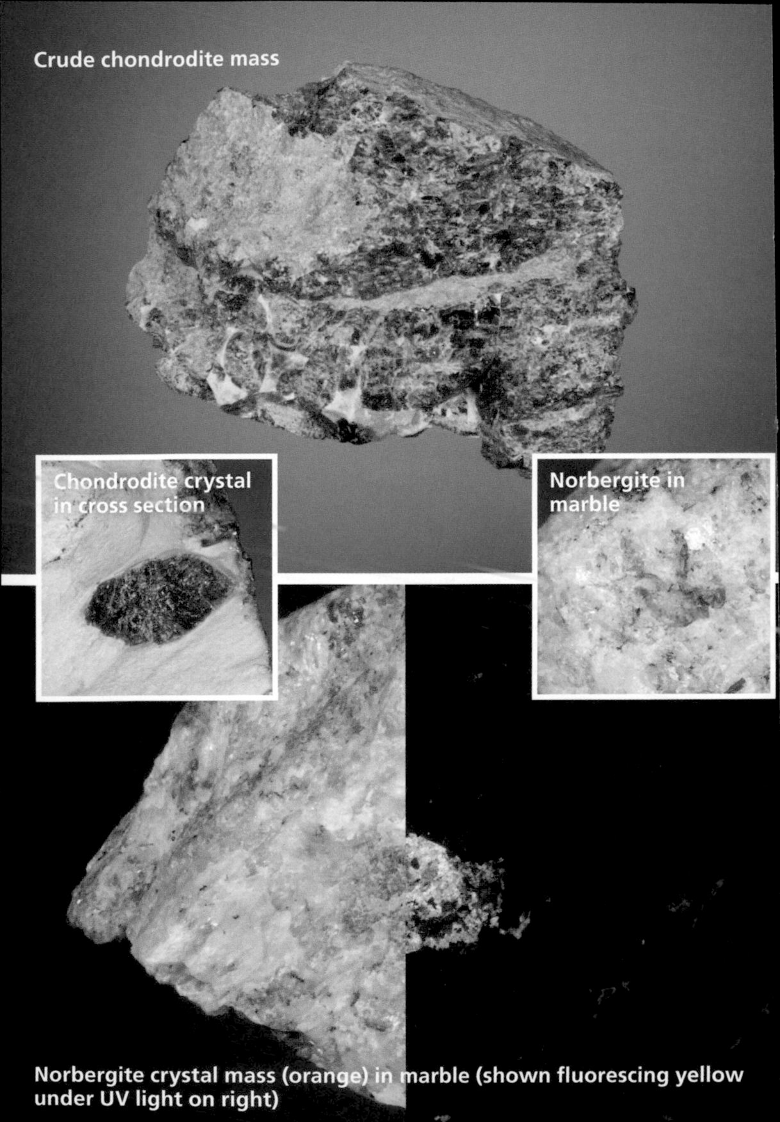

Crude chondrodite mass

Chondrodite crystal in cross section

Norbergite in marble

Norbergite crystal mass (orange) in marble (shown fluorescing yellow under UV light on right)

Humite group

HARDNESS: 6-6.5 **STREAK:** Gray

Primary Occurrence

ENVIRONMENT: Mines, quarries, outcrops, hills

WHAT TO LOOK FOR: Isolated, glassy, translucent, reddish crystals embedded in metamorphic rocks, often with diopside

SIZE: Crystals are smaller than a thumbnail, but masses may reach an inch or more

COLOR: Brown to reddish, yellow to orange or brown

OCCURRENCE: Rare

NOTES: The humite group is a rare family of complex silicate minerals that contain metals like iron and/or magnesium combined with fluorine and hydrogen; they form primarily as grains or embedded crystals within metamorphic rocks that formed when molten rock contacted preexisting rocks. As such, chondrodite, the most common humite group member, and norbergite, another member present in New York, are most often found in marble or skarn, often alongside minerals like graphite, magnetite, or serpentines. Chondrodite is a popular collectible due to its gem-like translucent quality and often vivid coloration, appearing as complex ball-like or elongated crystals, but it is more abundant as irregular masses. It can easily be confused with garnets (page 115) due to their similar appearance and host rock, but garnets are far more common and typically harder. Norbergite is quite rare and is virtually always seen as irregular masses in marble; a shortwave UV lamp will aid identification, as norbergite often fluoresces yellow.

WHERE TO LOOK: The Tilly Foster Iron Mine in Putnam County has produced world-class specimens, but accessing them today is impossible for amateurs. Numerous localities in Orange County, near Amity, Edenville and Pine Island have produced specimens embedded in marble from quarries.

Pyrobitumen (black) with dolomite (tan) in limestone cavity

Asphaltum (brown) on fluorite crystal

Asphaltum (brown) on fluorite crystal

"Herkimer diamond" quartz containing globular formation of pyrobitumen within

Hydrocarbons

HARDNESS: N/A **STREAK:** N/A

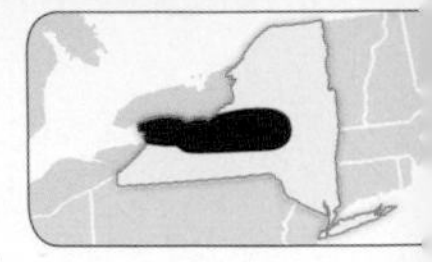
Primary Occurrence

ENVIRONMENT: Mines, quarries, outcrops, lowlands

WHAT TO LOOK FOR: Black to brown fluid-like coatings on rock or blob-like inclusions in minerals

SIZE: Hydrocarbon formations can vary greatly, but they are often smaller than your thumbnail

COLOR: Typically black, but also light to dark brown

OCCURRENCE: Uncommon

NOTES: Because of their largely organic origins, limestone and dolostone (page 149) have the potential in many geological environments to contain hydrocarbons, which are petroleum or oil compounds that consist primarily of hydrogen and carbon that is derived from ancient organic material. Hydrocarbons are not minerals, but typically occur with or within minerals such as calcite and fluorite. Several hydrocarbons are present in New York, but pyrobitumen (called "anthraxolite" in older texts) and asphaltum are most prominent. Pyrobitumen often has a lustrous wet, fluid appearance despite actually being hard and brittle, while asphaltum is lighter in color and typically sticky and greasy to the touch. While asphaltum is typically found as a coating or stain on minerals and isn't very collectible in itself, pyrobitumen can exhibit beautiful globular forms, most famously as rather spectacular inclusions within New York's "Herkimer diamonds" (page 133). Both can also have a pungent smell; combined with their other peculiar and unique traits, this makes pyrobitumen and asphaltum easy to identify.

WHERE TO LOOK: The "Herkimer diamond" mines in Middleville and St. Johnsville have produced specimens of pyrobitumen encased within the quartz. The Walworth Quarry in Wayne County produces fluorite specimens coated in asphaltum.

Crust of tiny hydromagnesite crystals (white) on serpentine

Detail of tiny crystals (under $\frac{1}{16}$") clustered on serpentine

Hydromagnesite

HARDNESS: 3.5 **STREAK:** White

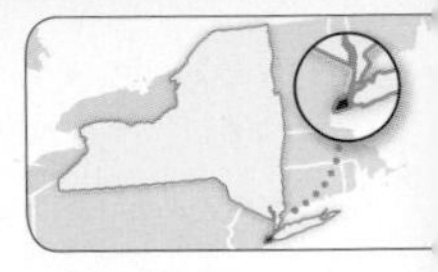

Primary Occurrence

ENVIRONMENT: Mines, outcrops

WHAT TO LOOK FOR: Chalky white crusts on serpentine, often with artinite

SIZE: Crystal mats can be up to palm-sized and rarely larger

COLOR: Typically white, sometimes colorless

OCCURRENCE: Very rare, no longer possible to collect

NOTES: Like artinite (page 53), with which it frequently occurs, hydromagnesite is a rare New York mineral that develops as an alteration product of serpentine minerals (page 199); this means that it develops when serpentines deteriorate. Found in only a few places on Staten Island, it was temporarily unearthed when housing construction exposed serpentine formations, making collecting possible for a short time. Hydromagnesite is a soft, delicate, white mineral that develops as tiny needle-like crystals, typically intergrown in countless numbers and occurring as a chalky mat or crust. Artinite grows in a very similar way, but its tiny, slender crystals tend to be longer, better developed, and are often glassier in luster. In addition, if a hydromagnesite crystal has grown large enough or is viewed under magnification, it's clear that its crystals are more broad than artinite's, with more of a blade-like appearance. But despite its rarity, the most classic appearance for New York's hydromagnesite is as an understated base from which more collectible artinite crystals grow.

WHERE TO LOOK: New York's only locality is Staten Island, with the best site being the Spring Street Occurrence. The serpentine hill that produced specimens is still there today, but it is fenced-off and inaccessible to collectors, so purchasing a specimen is likely your only option.

Large ilmenite mass (approx. 3")

Heavy sand containing ilmenite and magnetite (as well as fragments of rocks containing both magnetic minerals) attached to a magnet

Ilmenite

HARDNESS: 5-6 **STREAK:** Brownish black

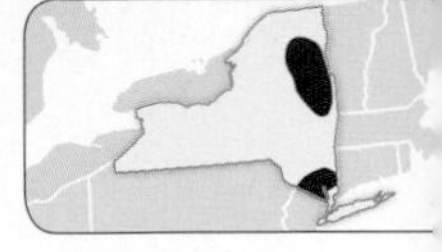

Primary Occurrence

ENVIRONMENT: Mountains, hills, mines, rivers, beaches, road cuts, outcrops

WHAT TO LOOK FOR: Weakly magnetic metallic blue-black grains or masses embedded in dark rocks or loose in sand

SIZE: Ilmenite crystals tend to be smaller than ¼ inch; masses may be fist-sized or larger

COLOR: Metallic black, occasionally with a bluish tinge

OCCURRENCE: Uncommon

NOTES: Ilmenite is an ore of titanium that was once mined in New York. It most often occurs as small grains or masses in igneous rocks such as gabbro. Metallic and black, often with a bluish sheen when viewed under bright light, ilmenite crystallizes as thin hexagonal (six-sided) plates which can resemble those of hematite (page 129). The most prominent different between the two, however, is ilmenite's weak magnetism. But while hematite isn't magnetic at all, magnetite (page 155) is, and when a specimen shows no crystal shapes, confusing the two is likely. Magnetite is very strongly magnetic, however, while ilmenite is only weakly so, and while it will bond to a magnet, it is often easily shaken free. Well-crystallized specimens are rare in the state but irregular grains or masses are more abundant, particularly as a component of "heavy sand," which is darker sand found in rivers and consists of dense minerals weathered from rocks. This material is commonly collected with a magnet.

WHERE TO LOOK: Ilmenite is widespread in New York but only in small amounts. The Newcomb area in Essex County once had enormous amounts of ilmenite, and fine crystals have been found in the Amity-Warwick mines in Orange County. Adirondack rivers will yield grains, collectible with a magnet.

Rough greenish jasper
Waxy luster
Jasper fragments from glacial gravel

Jasper

HARDNESS: ~7 **STREAK:** N/A

Primary Occurrence

ENVIRONMENT: All environments

WHAT TO LOOK FOR: Very hard, opaque masses, often with a smooth, waxy feel and appearance when weathered

SIZE: Masses can range greatly, but are usually smaller than a fist

COLOR: Varies greatly; generally gray to brown or reddish, but also yellow to green, or multicolored

OCCURRENCE: Common

NOTES: Jasper is popular because it has vibrant colors and is easy to identify. It is essentially a more colorful variety of chert (page 77) though its method of formation can differ. As such, it is a variety of microcrystalline quartz (page 185), consisting of tightly bonded microscopic grains of quartz, but it is frequently stained by iron-bearing minerals and other materials that give it its trademark colors. Like all other microcrystalline quartz, jasper has no structure of its own but instead takes on a shape dictated by its surroundings, and therefore often appears as irregular masses, nodules (round masses), or veins within other rocks. And like any form of quartz, jasper is very hard and exhibits conchoidal fracturing (when struck, circular cracks appear). When worn, it shows a smooth, wax-like appearance and texture, but is more dull, rough, and sharp-edged when freshly broken, all of which are traits shared by chalcedony (page 71), but chalcedony is always more translucent. Most specimens will be found loose in gravel where they were deposited by glaciers that carried them from the north.

WHERE TO LOOK: Jasper easily withstood the glaciers and as a result can be found in many places throughout the state, particularly in gravel-rich areas. Any major river, especially the Hudson and the Mohawk, will yield worn pebbles.

Labradorite masses
Thin parallel crystals
Rare orange schiller
Typical blue-green internal schiller seen at specific light angles

Labradorite

HARDNESS: 6-6.5 **STREAK:** White

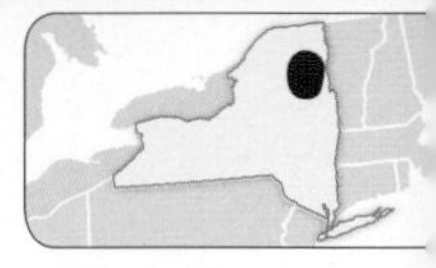

Primary Occurrence

ENVIRONMENT: Mountains, outcrops, road cuts, rivers, mines

WHAT TO LOOK FOR: Hard, blocky, dark mineral masses with internal flashes of blue-green

SIZE: Specimens can vary from tiny grains to fist-sized masses

COLOR: Gray to black, sometimes bluish or greenish; with a bright blue-green to yellow-orange internal schiller

OCCURRENCE: Uncommon

NOTES: Labradorite is among the most impressive of the feldspars (page 93) and is widely collected and prized for its beautiful blue internal schiller, or reflections, which are visible when it is rotated under bright light. Technically a variety of anorthite, a feldspar in the plagioclase group, labradorite generally occurs in darkly colored blocky masses that can be identified using the same criteria as any other feldspar. It also yields a beautiful play of light when oriented in just the right way, and this occurs because of the unique way its crystals developed. A mass of labradorite isn't a single feldspar crystal. Instead, it's actually many thin crystals that have grown tightly together and are oriented parallel to one another. (This can be observed on some specimens with striated, or grooved, surfaces.) As light enters a specimen, it bounces between the individual crystals and refracts back to our eye in a dazzling display of color, a phenomenon known as labradorescence. Labradorite can be confused with moonstone (page 219), which also has a schiller, but moonstone is a much lighter colored feldspar and its "flash" is more white.

WHERE TO LOOK: Look in the Adirondack Mountains, particularly in Essex County. The peaks and rivers around Lake Placid and Saranac Lake yield specimens weathered from anorthosite.

Rough mass of lapis lazuli with lazurite (blue), calcite (white), and pyrite (brassy)

Sawn specimen showing desirable vivid blue color

Lapis Lazuli/Lazurite (Haüyne)

HARDNESS: 5-5.5 **STREAK:** Blue

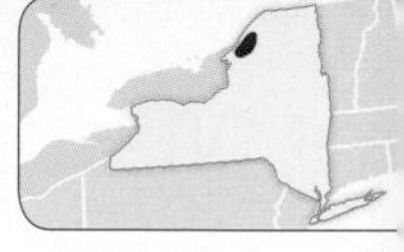

Primary Occurrence

ENVIRONMENT: Mines

WHAT TO LOOK FOR: Rock with vividly blue coloration, also containing white and brassy spots

SIZE: Specimens are typically smaller than a foot long

COLOR: Multicolored; primarily light to dark blue, with white and brassy spots

OCCURRENCE: Very rare; no longer collectible

NOTES: The ancient Persians called it "lazhward" which was later modified into Latin as *lazulum*, from which we get our modern word azure; for thousands of years, lazuli has been synonymous with the color blue, combined with "lapis," Latin for stone. Lapis lazuli, the "blue stone," is most famously mined in Afghanistan but is also found in a handful of other places, one of which is the Balmat-Edwards area of New York. When you see it, you won't mistake it for anything else. Lapis, as it is more commonly known, is actually a very rare rock consisting primarily of lazulite, the name for the vivid blue variety of the mineral haüyne, along with some white calcite and brassy pyrite mixed in. It forms within metamorphic rocks such as marble, under a specific set of circumstances, and unfortunately it is so rare and highly prized that collecting your own specimen is nearly impossible today. Still, specimens are often available for sale, and are identified by sight alone, as virtually nothing else in the state has the same intense blue coloration and is commonly associated with pyrite.

WHERE TO LOOK: Only the Balmat-Edwards Zinc District in St. Lawrence County has produced lapis in New York, and primarily only from the St. Joe and Edwards Mines, both of which are closed to casual collectors.

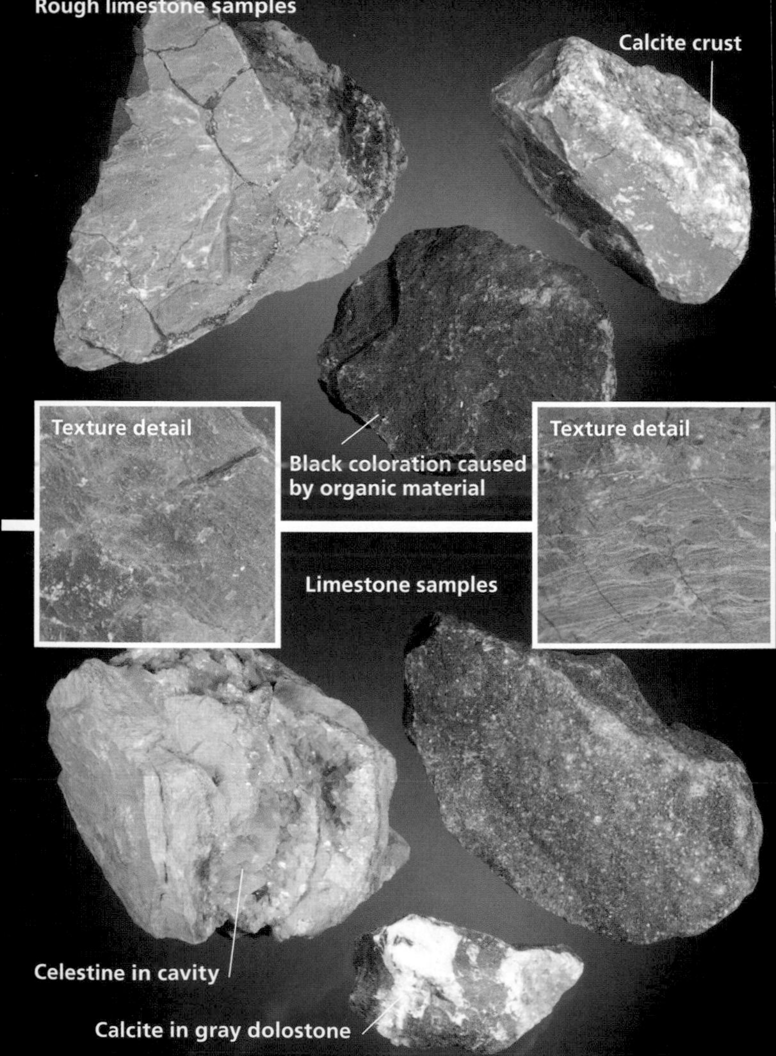

Rough limestone samples
Calcite crust
Texture detail
Black coloration caused by organic material
Texture detail
Limestone samples
Celestine in cavity
Calcite in gray dolostone

Limestone/Dolostone

HARDNESS: 3-4 **STREAK:** N/A

Primary Occurrence

ENVIRONMENT: Lowlands, hills, quarries, mines, rivers, outcrops, road cuts

WHAT TO LOOK FOR: Soft yet tough, compact light-colored rock often with a chalky feel, abundant in flatter regions

SIZE: As a rock, limestone can be found in any size

COLOR: White to tan or brown, also frequently light to dark gray

OCCURRENCE: Very common

NOTES: Limestone is one of the most common sedimentary rocks in New York and therefore is also one of the easiest to find and identify. It formed at the bottoms of shallow marine seas when the remains of hard-shelled organisms settled into thick beds. Originally these organic sediments consisted of aragonite (page 49), an unstable mineral that in time converted into calcite (page 63). In the process, it recrystallized and eventually cemented together to form limestone. Though limestone contains some dolomite, clay, and minor amounts of quartz, it consists of over 50 percent calcite, making it easy to scratch with a knife and easy to dissolve in even weak acids; undiluted vinegar will make it effervesce, or fizz. Typically white or tan but darker when containing more organic material, such as oil, limestone develops a chalky texture and a white color when worn; it also often contains fossils. If the fossils have since dissolved, cavities called vugs are left behind, often containing calcite or dolomite crystals. Finally, a dolomite-rich variety called dolostone is prevalent in the region but generally visually identical.

WHERE TO LOOK: Much of the state is covered by limestone and dolostone. The Catskills are composed largely of these stones, as is most of the bedrock in the western and southern portions of the state; Niagara Falls crash over limestone cliffs.

Löllingite masses and grains (silvery) throughout metamorphic rock

Löllingite mass

Silvery löllingite mass embedded in metamorphic rock

Löllingite

HARDNESS: 5-5.5 **STREAK:** Grayish black

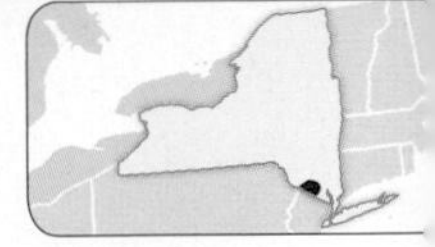

Primary Occurrence

ENVIRONMENT: Mines

WHAT TO LOOK FOR: Brittle and lustrous silvery metallic mineral, often as elongated crystals, embedded in metamorphic rock

SIZE: Specimens are typically only an inch or two in size

COLOR: Metallic gray to silvery white

OCCURRENCE: Very rare

NOTES: Löllingite, also often spelled "loellingite," is among the rarest minerals in New York, with only one small area near the New Jersey border having ever produced it. It is a dense, typically silvery metallic mineral that consists only of iron and arsenic (it's generally safe to handle, but wash your hands thoroughly afterwards). It is found in the metamorphic rocks of the Franklin Marble Formation, which extends into New Jersey, and other rare and unusual minerals are found in New Jersey's part of the geological formation. When well-formed, it's easy to note its elongated, prismatic crystals with typically diamond-shaped cross sections, but in New York it is generally only seen as masses or grains within its host rock, with little discernible shape. This makes it very easy to confuse with arsenopyrite (page 51) with which it frequently occurs. Both have a similar color, hardness, and streak, so if no crystals are present it's virtually impossible for amateurs to differentiate the two. But its attractively bright luster and its rarity make it a popular New York collectible for those persistent enough to seek it out.

WHERE TO LOOK: Only mines in and around Warwick in Orange County, near the New Jersey border, have ever produced this extremely rare mineral in New York. Finding your own specimen is highly unlikely, not only because of private land, but because most of the old mines and their dumps are grown-over and inaccessible.

Cluster of cubic magnetite crystals from Balmat-Edwards

Magnetite cube

Magnetite-rich sand on magnet

Octahedral magnetite crystals in schist

Magnetite

HARDNESS: 5.5-6.5 **STREAK:** Black

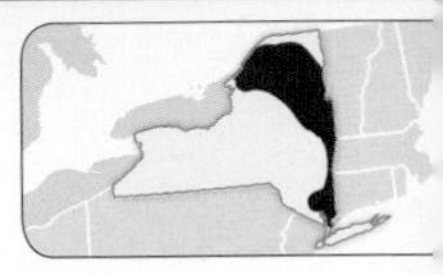

Primary Occurrence

ENVIRONMENT: Mountains, hills, mines, rivers, beaches, road cuts, outcrops

WHAT TO LOOK FOR: Metallic black pyramid-shaped crystals or masses that are attracted to a magnet

SIZE: Individual crystals are rarely larger than an inch while clusters can be fist-sized or larger

COLOR: Metallic black; sometimes with rust-colored surfaces

OCCURRENCE: Common

NOTES: A common iron ore and the most abundant mineral of the spinel group (page 215), magnetite is a curious mineral that is both relatively easy to find and identify. It is always metallic black in color (unless weathered, in which case it will have a rusty surface coating) and can frequently be found crystallized, typically as octahedrons, which are eight-faced shapes resembling two pyramids placed base-to-base, but more rarely as cubes, which are exclusive to one mine on Earth: the ZCA No. 4 Mine in the Balmat-Edwards area. While the lustrous cubic crystals are legendary, the mines are closed and crystals aren't obtainable by casual collectors; octahedral crystals, grains, and irregular masses are more likely finds and typically embedded in rocks like schist, gneiss, and granite. Aside from its hardness and appearance, magnetite's most diagnostic identifying trait is its namesake magnetism, and it will strongly bond to a magnet. This makes it only confused with ilmenite (page 141), which is less common and only weakly bonds to a magnet.

WHERE TO LOOK: Schists in the Adirondacks, particularly in Warren and Essex Counties, have yielded octahedral crystals, and any sand or gravel in the state has the potential to produce magnetic grains or even loose crystals.

Very pure, lustrous, calcite-rich marble
Texture detail
Individual calcite crystals
Masses of freshly broken marble
Mica crystals
(brown)
Diopside masses (gray)

Marble

HARDNESS: ~3 **STREAK:** N/A

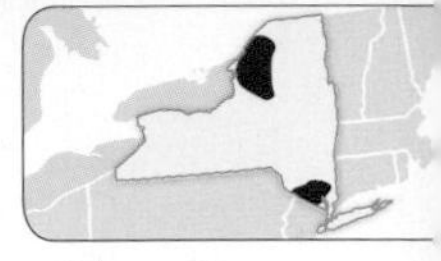
Primary Occurrence

ENVIRONMENT: Lowlands, hills, quarries, mines, outcrops, road cuts, rivers

WHAT TO LOOK FOR: Light-colored rock with a lustrous crystalline texture, rich with calcite

SIZE: As a rock, marble can be found in any size

COLOR: White to gray; yellow to brown if impure, and sometimes mottled with darker spots

OCCURRENCE: Common

NOTES: Used for millennia in sculptures, marble is a metamorphic rock composed largely of calcite (page 63), and it is a rock in which many of New York's most collectible minerals form. Marble forms when limestone is buried deep within the earth and exposed to the metamorphic forces of heat and pressure. This causes its tiny calcite grains to combine and recrystallize in a larger size, growing into a dense, compact mass of interlocking coarse crystalline grains. The other minerals within limestone, such as dolomite and clay, are also often transformed, developing into different minerals, sometimes highly collectible or rare ones. Marble's overall color is generally white to gray and its texture, hardness, and luster are distinctly like that of calcite, its predominant constituent, but the other minerals that may form within it, like diopside or chondrodite, can give it a mottled appearance or change its texture slightly. Whatever its appearance, marble generally isn't difficult to identify; if in doubt, undiluted vinegar will make it fizz strongly. Compare with skarn on page 209.

WHERE TO LOOK: Orange County is home to a portion of the famous Franklin Marble Formation. The Amity-Warwick area has produced huge amounts of marble that bear rare, collectible minerals, both from quarries and public outcrops.

Fine cluster of bladed marcasite crystals within calcite geode

Needle-like crystals

Bladed crystal

Tiny (⅛") marcasite needles on dolomite crystals

Marcasite

HARDNESS: 6-6.5 **STREAK:** Dark gray to black

Primary Occurrence

ENVIRONMENT: Lowlands, hills, mountains, mines, quarries, road cuts, outcrops

WHAT TO LOOK FOR: A hard, brassy metallic mineral occurring as plate- or blade-like crystals in limestone

SIZE: Crystals are typically thumbnail-sized or smaller

COLOR: Pale brass-yellow to metallic gray or brown, sometimes with multicolored surface coating

OCCURRENCE: Common

NOTES: Marcasite consists of iron and sulfur, just like pyrite (page 177), but the two minerals differ in crystal structure because conditions varied when they formed. So while they both exhibit similar hardness and streak colors, marcasite does not share pyrite's cubic crystals and instead appears more tabular (thin, plate-like) and is often intergrown in clusters with multiple serrated points and deep striations (grooves). Marcasite can also form as tiny, thin blades or needles, often on calcite or dolomite, and these can be identified by color and their generally diamond-shaped cross section as well as their primary occurrence in cavities within limestone. In fact, all varieties of marcasite typically formed in limestone or other sedimentary rocks, especially more massive specimens which can be distinguished from pyrite by their paler color. Chalcopyrite (page 75) is also similar but much softer; pyrrhotite (page 183) has a similar color but is rarer and magnetic. Lastly, marcasite concretions (ball-like formations) can be found in sedimentary rocks as well and are rough and crumbly.

WHERE TO LOOK: Central New York quarries, particularly in Montgomery County, have yielded the best specimens in the state, but most of the counties in the Erie-Ontario Lowlands (along which I-90 runs) have produced the mineral.

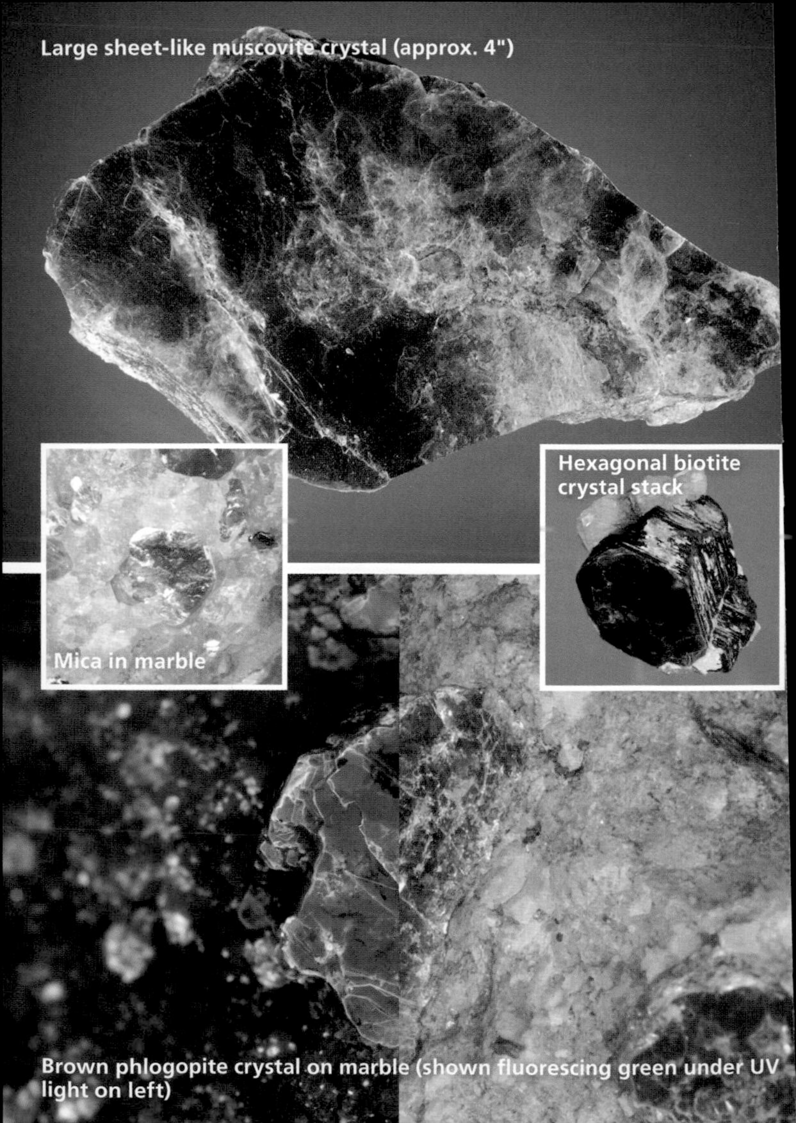

Large sheet-like muscovite crystal (approx. 4")

Mica in marble

Hexagonal biotite crystal stack

Brown phlogopite crystal on marble (shown fluorescing green under UV light on left)

Mica group

HARDNESS: 2.5-3 **STREAK:** Colorless

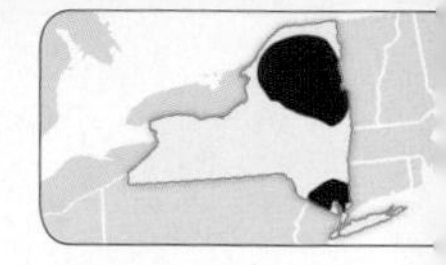

Primary Occurrence

ENVIRONMENT: All environments

WHAT TO LOOK FOR: Lustrous, generally dark-colored minerals that appear as thin, flexible sheets, often embedded in rocks

SIZE: Crystals are often paper-thin and smaller than a thumbnail; rare specimens are palm-sized or larger

COLOR: Gray to yellowish or brown; less commonly black or green; rarely blue

OCCURRENCE: Very common

NOTES: The mica group is a family of minerals that all share a set of unique and easily identifiable traits. All of New York's micas form as thin, very flexible, hexagonal (six-sided), sheet-like crystals that grow in flaky, layered stacks called books. Examples include muscovite (the most common), phlogopite, biotite, and clintonite, among others. In most cases, individual page-like crystals can be peeled off the rest of the stack. These brightly lustrous, sometimes seemingly metallic crystal groups are most often seen in coarse-grained igneous rocks, especially granite (page 119), but less finely formed examples are more common. That's because the micas are primarily rock-builders, meaning they are most abundant as a constituent of rocks. In many rocks, the micas are nearly microscopic and may only be visible as "glitter," while some metamorphic rocks, such as some schists (page 195), consist largely of compacted micas. Micas are best identified by the traits mentioned above, but also by their low hardness, which is higher than that of similarly shaped graphite (page 121), but softer than other hexagonal minerals.

WHERE TO LOOK: Micas are found easily in most non-sedimentary environments, particularly in waterworn pebbles of granite or schist. The Adirondacks are prime territory for specimens.

Blue phlogopite
Clintonite (brown)
"Gieseckite" crystal
in calcite

Mica group, varieties

HARDNESS: 2.5-3 **STREAK:** Colorless

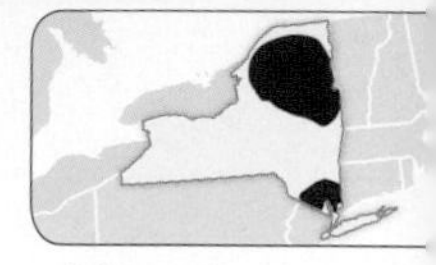

Primary Occurrence

ENVIRONMENT: Mines, quarries, outcrops, mountains, hills

WHAT TO LOOK FOR: Lustrous, generally dark-colored minerals that appear as thin, flexible sheets, often embedded in rocks

SIZE: Crystals are often paper-thin and smaller than a thumbnail; rare specimens are palm-sized or larger

COLOR: Gray to yellowish or brown; rarely black, green, or blue

OCCURRENCE: Rare to very rare, depending on variety

NOTES: Mica minerals form in a wide range of environments, and though some only form in specific rock types, others, like muscovite, are able to form in dozens of different settings. Given this variety in the environments in which they form, many mica varieties exist, especially in New York. Some varieties are somewhat superficially different, and only vary in color. Phlogopite, normally gray or brownish, can be found in the state in beautiful shades of blue, making for rare and popular specimens. Other varieties are structurally different, such as with "gieseckite," an old term for a mixture of muscovite and illite (a clay-like mica) that has pseudomorphed (been chemically replaced while retaining the external shape) an unknown mineral, making it an interesting rarity. New York was also the site of the initial discovery of clintonite, a rare mica mineral. It forms exclusively in the marble of Orange County due to the unique geology of the region. It is bronze-colored and can be harder than typical micas, up to a 6 on the Mohs scale, but it is very rare.

WHERE TO LOOK: Clintonite is only found in the Warwick-Amity area. The famous blue phlogopite is found in the Talcville area mines; the only place on earth where you can find the incredibly rare gieseckite is near Natural Bridge in Lewis County.

Fine millerite cluster (approx. ½") on hematite-stained quartz

Millerite cluster (partially changed to pecoraite) in quartz

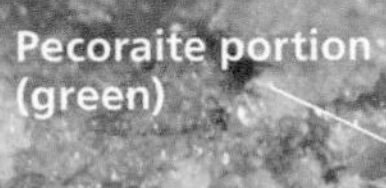

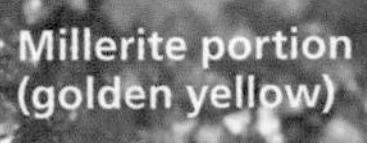

Millerite

HARDNESS: 3-3.5 **STREAK:** Greenish black

Primary Occurrence

ENVIRONMENT: Mines

WHAT TO LOOK FOR: Long hair- or needle-like brassy crystals, often in divergent groupings referred to as "sprays"

SIZE: Single crystals are rarely longer than ½ inch

COLOR: Brass-yellow, sometimes with a greenish tinge

OCCURRENCE: Very rare

NOTES: A combination of nickel and sulfur, millerite's striking crystals make it among the most interesting metallic minerals, and one of the easiest to identify. Unfortunately, it is also among the rarest in New York. Its crystals develop as brassy yellow, extremely fine, delicate needles often clustered together in divergent groups or as tangled mats nestled within cavities in rocks, particularly iron-rich ores in New York. Millerite's hair-like crystals are flexible and will bend before they break, which, in combination with their color, luster, form, and rarity, is typically enough for identification. The only mineral that will resemble it in New York is pecoraite (page 173), which is actually a pseudomorph, or chemical alteration, of millerite that retains millerite's crystal shape, but pecoraite is green, which is the key difference. In New York, millerite is only found in reddish hematite (page 131), often with siderite (page 203). Specimens are highly desirable and valuable, so even if one must be purchased, a specimen is worth pursuing.

WHERE TO LOOK: Millerite is only found at one significant site in New York State: the Sterling Mine in Antwerp. This old mine, closed in 1910, was the first locality in the U.S. where millerite was identified, and has produced some of the finest specimens in the world. The dumps still yield occasional specimens, if visited with permission or with a rock club.

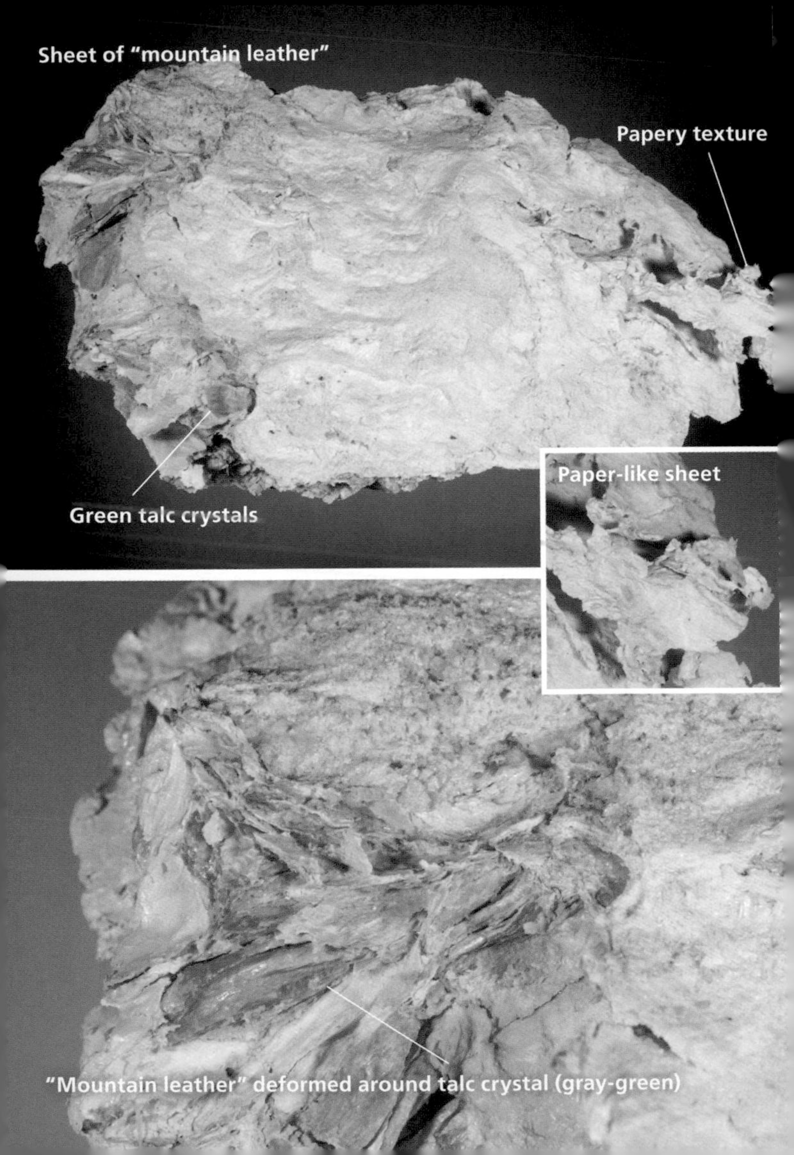
Sheet of "mountain leather"
Papery texture
Green talc crystals
Paper-like sheet
"Mountain leather" deformed around talc crystal (gray-green)

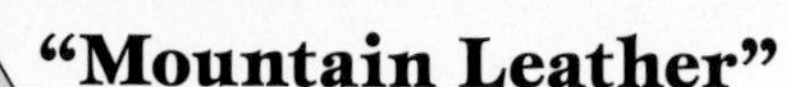

"Mountain Leather"

HARDNESS: <5 **STREAK:** White

Primary Occurrence

ENVIRONMENT: Mines

WHAT TO LOOK FOR: Unusually flexible, fabric-like sheets or masses of friable, fibrous material within metamorphic rocks

SIZE: Mountain leather is found as palm-sized sheets or larger

COLOR: White to gray, yellow to brown

OCCURRENCE: Very rare

NOTES: "Mountain leather," also called "mountain paper" or "mountain wood," is known by a number of old collector's names thanks to its unique properties. At first glance, it looks like it consists of woven organic fibers, as it's flexible and able to be pulled apart or cut into pieces. Nonetheless, this strange material is actually a crystallized mineral. To be specific, it is a kind of asbestos, or a mineral formation consisting of countless tiny fibers, which can, in the right conditions, become airborne and pose a health risk. A respirator and some gloves are an easy precaution, and are recommended. Several minerals can develop as asbestos; in New York, they include tremolite (page 229), actinolite (page 37), palygorskite, and chrysotile (page 199), and they sometimes form in combination with each other. They took on this interesting form when they developed within crevices in metamorphic rocks, particularly marble, when other minerals, such as serpentines, were chemically altered. With its peculiar appearance and flexibility, combined with its association with other soft minerals, like talc and serpentines, you're unlikely to confuse mountain leather with anything else. Unfortunately, it's rare enough that you won't likely find it without expert help.

WHERE TO LOOK: Mines in Putnam and Dutchess Counties are about the only places it has turned up in New York and it's rare enough that specimens aren't often available for sale.

Rough mudstone
Fairly even coloration
Mudstone detail
River-worn siltstone
Extremely fine-grained texture
Limonite-rich samples
Siltstone

Mudstone/Siltstone

HARDNESS: N/A **STREAK:** N/A

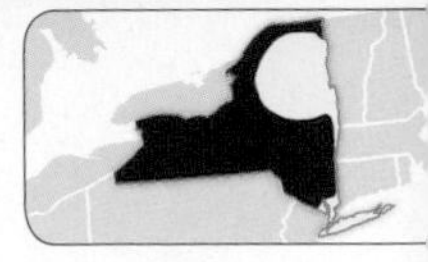

Primary Occurrence

ENVIRONMENT: Lowlands, hills, quarries, road cuts, outcrops, rivers

WHAT TO LOOK FOR: Soft, dense rocks that resemble hardened mud and consist of nearly microscopic grains

SIZE: As rocks, mudstone and siltstone can occur in any size

COLOR: Light to dark gray, tan to brown, yellowish, occasionally reddish

OCCURRENCE: Common

NOTES: Mudstone and siltstone are two sedimentary rocks that consist of extremely small mineral particles that have weathered; siltstone consists of silt-sized particles, around 1/5,000 of an inch in size, and mudstone of mud-sized particles, measuring only 1/12,500 of an inch. While the grains are clearly too small to see without powerful magnification, what this means for rockhounds is that both rocks appear extremely fine-grained, even-colored, and are fairly soft, often with a slightly gritty feel. Mudstone is quite similar to shale (page 201); that's because mudstone is essentially the same thing as shale, but lacks shale's bedded layers. Telling mudstone and siltstone apart is beyond the means of most amateurs, but a general way to distinguish them is that mudstone contains large amounts of clay, which makes it a bit softer and makes it split apart more easily than siltstone. Either type of rock may contain concretions (page 81) and, less commonly, fossils.

WHERE TO LOOK: Any low-lying area rich in shales will also yield some amount of mudstone or siltstone; this includes the entire southern portion of the state, the Catskills, and especially the far western portion of the state.

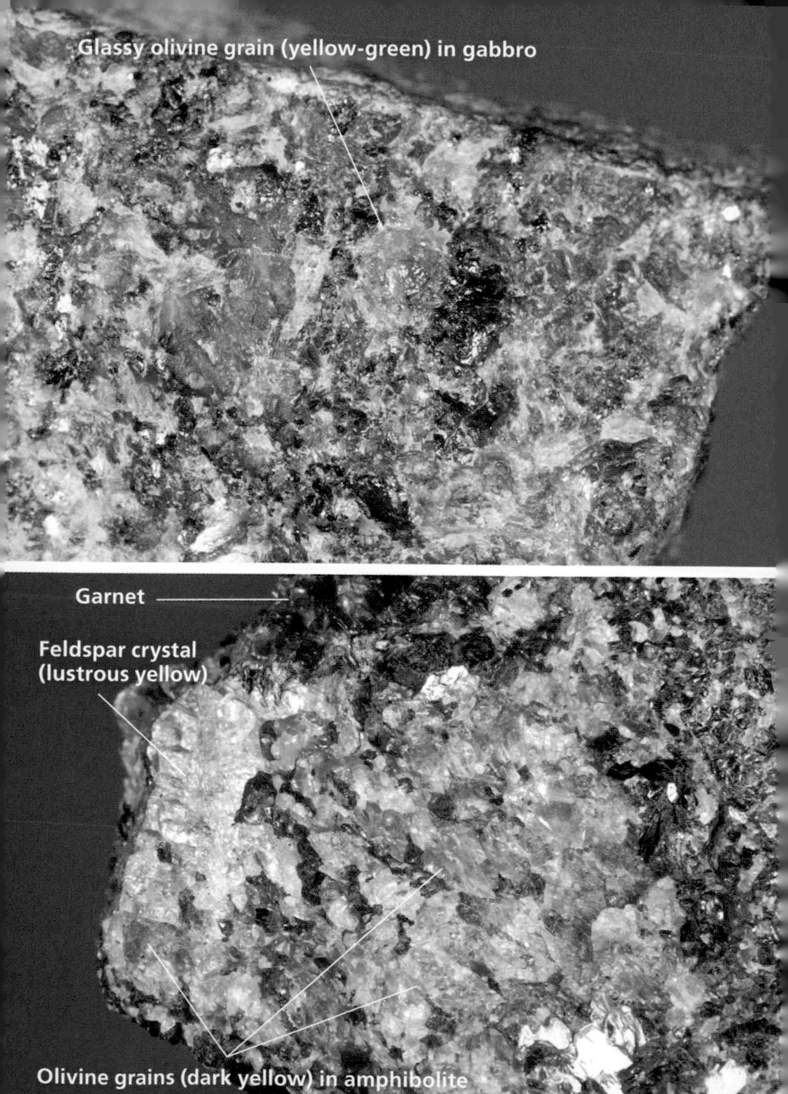
Glassy olivine grain (yellow-green) in gabbro
Garnet
Feldspar crystal
(lustrous yellow)
Olivine grains (dark yellow) in amphibolite

Olivine group

HARDNESS: 6.5–7 **STREAK:** Colorless

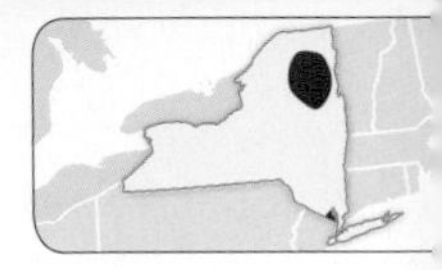

Primary Occurrence

ENVIRONMENT: Mountains, hills, mines, rivers, beaches, outcrops, road cuts

WHAT TO LOOK FOR: Very hard, glassy, green-yellow grains embedded in dark rocks

SIZE: Most olivines are found as tiny grains no more than ¼ inch

COLOR: Yellow to green or dark green, yellow-green; less commonly brown

OCCURRENCE: Common

NOTES: The olivine group is a small but important family of rock-builders, or minerals that are most prevalent as a constituent of rocks, particularly dark igneous rocks like diabase and gabbro. Forsterite, the most common olivine mineral, can frequently be seen as irregularly shaped, glassy, yellow-green masses or grains in gabbro (page 111); though it can be difficult to pick out alongside other dark minerals. Forsterite's high hardness, luster, and typical translucency in bright light help differentiate it. Few localities in New York produce very collectible specimens, and crystals aren't found in the state, but it is still common enough in certain settings that collectors should be able to recognize it. For example, in river sand or gravel, loose masses or grains may be found weathered free from their host rock, and though tiny, specimens can be colorful and almost gem-like, making them easy and rewarding finds. Epidote (page 91) may share a similar color and hardness, but it is less abundant and forms as veins or crusts rather than as embedded grains.

WHERE TO LOOK: The igneous rocks of the Adirondacks contain olivine minerals as tiny embedded grains, while river gravel and sand in the region can yield small freed masses.

Very fine pecoraite cluster (approx. ¼") in quartz
Pecoraite (⅛")
Pecoraite cluster (green) in hematite (red)
Metallic hematite

Pecoraite

HARDNESS: 2.5 **STREAK:** Pale green

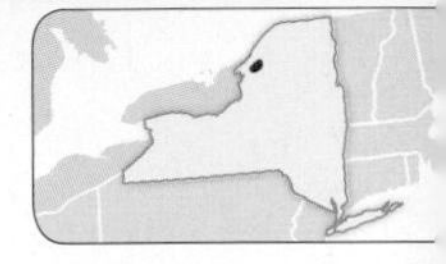

Primary Occurrence

ENVIRONMENT: Mines

WHAT TO LOOK FOR: Tiny green needle-like crystals arranged into radial clusters within iron-rich rock

SIZE: Crystals are tiny, rarely larger than ¼ inch

COLOR: Light green, yellow-green

OCCURRENCE: Very rare

NOTES: One of New York's most beautiful and interesting rarities, pecoraite is a nickel-bearing silicate belonging to the serpentine group (page 199). While in other parts of the world it forms as masses and veins, in New York it is found only in a single mine where it develops exclusively as a pseudomorph of millerite (page 165). A pseudomorph is a mineral formation that is the result of another mineral undergoing a chemical change but retaining its outward appearance; a pseudomorph therefore is a mineral with the crystal shape of a completely different mineral. In pecoraite's case, it retains the radially arranged needle-like appearance of the original millerite and is found tucked within vugs (irregular cavities) alongside siderite and quartz in the hematite-rich rock from the mine. Pecoraite is easily distinguished from millerite, though, thanks to its notable difference in color. While millerite is a brass-yellow and metallic, pecoraite is green, and dull to greasy in luster. Nothing else in the state resembles it.

WHERE TO LOOK: Only a single locality for pecoraite exists in New York: the Sterling Mine in Antwerp. Though long-closed, the mine has produced some of the world's best specimens; the dumps may still occasionally yield specimens, but most collectors will have to purchase one.

Masses of pegmatite containing collectible minerals

Pegmatite exhibiting feldspars (tan), quartz (white), and titanite (dark brown)

Pegmatite

HARDNESS: N/A **STREAK:** N/A

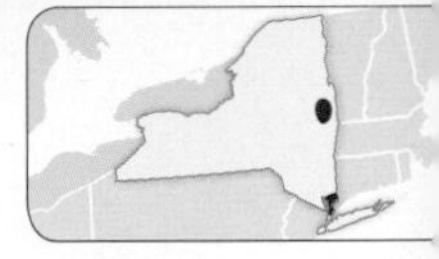
Primary Occurrence

ENVIRONMENT: Hills, mountains, quarries, outcrops

WHAT TO LOOK FOR: Extremely coarse-grained rock with large, visible crystals, containing lots of quartz and feldspar

SIZE: As a rock, pegmatite can be found in any size

COLOR: Multicolored; mottled in mostly white, tan to brown, pink, and gray to black

OCCURRENCE: Uncommon

NOTES: When granite (page 119) is still magma (molten rock) deep within the earth, the warmth of the inner earth and the insulating nature of the surrounding rock means that it cools very slowly. This allows the minerals within it a long period of time to crystallize, which lets them reach the coarse, visible grain size we see in granite today. But at the bottom of a body of magma, where the earth is even hotter, the minerals crystallize even more slowly, taking millennia to fully form. This part of a granite formation is called pegmatite, and is the coarsest, most well-developed form of granite. Many rare, unusual minerals form exclusively in pegmatite, such as rose quartz (page 187), but even the common minerals within it may be spectacularly large or well-formed. The primary minerals you'll see in pegmatite are quartz (page 185) and feldspars (page 93), and the rock is easily identified by sight alone, as nothing else will be quite so coarse-grained.

WHERE TO LOOK: There aren't many significant pegmatite outcrops in New York, but Saratoga County is home to some prolific quarries which may be searchable with permission, particularly in the Greenfield area. Westchester County is also a source of pegmatite material, though urban development has eliminated some sites.

Samples of pyrite
Limonite coating (brown)
Attached schist
Cubic crystals
Complex crystal
Pyritohedron on dolomite
Pyrite cube (brassy) in schist (dark green)

Pyrite

HARDNESS: 6-6.5 **STREAK:** Greenish black

Primary Occurrence

ENVIRONMENT: Lowlands, hills, mountains, mines, quarries, road cuts, outcrops

WHAT TO LOOK FOR: Hard, metallic brass-yellow masses or cubic crystals embedded in rock

SIZE: Crystals are rarely larger than an inch, masses may be larger

COLOR: Brass-yellow to metallic brown; rusty when weathered

OCCURRENCE: Common

NOTES: A relatively easy metallic mineral to find and identify, pyrite is abundant and popular. Like marcasite (page 159), pyrite is iron sulfide, a chemical combination that forms in a wide range of geologic environments, particularly in sedimentary rocks like shale and limestone or in metamorphic rocks like schist. Long known as "fool's gold" due to its typical brass-yellow color, pyrite can also develop a rusty orange or brown surface coating when weathered. Crystals are not uncommon and are generally cubic, often intergrown and with striated (grooved) sides, but rarer shapes exist in New York, such as "pyritohedrons," a shape that exhibits twelve five-sided faces, and octahedrons, which resemble two pyramids placed base-to-base. Irregular masses, often with a rough and ragged texture, are also abundant. Whatever its form, pyrite can be distinguished from similar looking chalcopyrite (page 75), and pyrrhotite (page 183) by its hardness—pyrite is harder than them—and from marcasite by its color, as marcasite is more gray.

WHERE TO LOOK: Anywhere there is exposed sedimentary or metamorphic rock will be a good place to start. The Pierrepont area mines, road cuts in Onondaga County, and quarries in Niagara County have all produced fine crystals.

Pyrite concretions
Pyritized ammonite fossil
Cluster of pyrite crystals

Pyrite, varieties

HARDNESS: 6-6.5 **STREAK:** Greenish black

Primary Occurrence

ENVIRONMENT: Lowlands, hills, mines, quarries, road cuts, outcrops

WHAT TO LOOK FOR: Hard, metallic brass-yellow clusters with rough surfaces, loose or in rock

SIZE: Crystals are rarely larger than an inch, while masses can be fist-sized

COLOR: Brass-yellow to metallic gray or brown; sometimes with a rusty coating

OCCURRENCE: Uncommon to very rare, depending on variety

NOTES: Pyrite can form in such a variety of environments that many interesting types of specimens can be found, especially in New York. Concretions are one of the most abundant varieties of pyrite; these nodules or rounded cluster-like formations form in sedimentary rocks, particularly shale (page 201) and developed within cavities left behind by fossils whose organic material reacted with seawater, forming the pyrite. Concretions and their globular shapes typically are more grayish in color than typical pyrite and have rough, ragged surfaces in which numerous little crystal shapes can often be seen. These aren't difficult to identify, though marcasite (page 159) can form similar concretions, but they are unstable and can crumble to dust in collections. But while the original fossil's shape and features are largely lost in pyrite concretions, pyritized fossils are another story. In these specimens, the hard parts (and in very rare cases, even the soft parts) of marine animals are well preserved as pyrite, making for stunning and sought-after specimens.

WHERE TO LOOK: Erie County has produced large amounts of concretions, particularly in the Spring Creek area. Impressive pyrite-replaced fossils are rare, but are found near Alden.

Masses of various pyroxene minerals
Enstatite (brown)
Augite (gray-green)
Enstatite crystal mass
Black pyroxene in gabbro
Fine enstatite crystal (black) in Gore Mountain garnet (red)

Pyroxene group

HARDNESS: 5-6.5 **STREAK:** White to gray-green

Primary Occurrence

ENVIRONMENT: All environments

WHAT TO LOOK FOR: Hard, dark, glassy masses, grains, or blocky elongated crystals embedded within coarse-grained rocks

SIZE: Most pyroxenes occur as crystals no larger than an inch or two; masses may be up to fist-sized

COLOR: Dark brown to gray or black, light to dark green

OCCURRENCE: Very common

NOTES: Closely related to the amphibole group (page 39), the pyroxenes are also an important family of rock-builders, or minerals that are most common as constituents of rocks, particularly dark igneous rocks like basalt. It's a large group, and several pyroxenes are present in New York, particularly augite and diopside (page 85), which are common. Enstatite and the ultra-rare donpeacorite (discussed on page 89) are also found here. Most take the form of glassy, black to green, blocky, generally elongated, prismatic crystals with angled tips, often within pegmatites (very coarse-grained granite outcrops) or metamorphic rocks, but these formed only under ideal conditions. Instead, it is more common to find pyroxenes as indistinct dark grains in rocks, though all share a key identifying trait: when broken, they cleave at nearly 90-degree angles, making for blocky, step-like breaks. Close observation of the shiny, dark spots in granite or gabbro will illustrate this characteristic. This will help tell them apart from amphiboles, which show a more fibrous or silky luster.

WHERE TO LOOK: Numerous localities for fine crystals are found in St. Lawrence County outcrops and mines as well as in Orange County near Warwick, Amity, and Edenville. The mountains of Essex County and the rest of the Adirondacks yield endless samples in the rocks that make up the peaks.

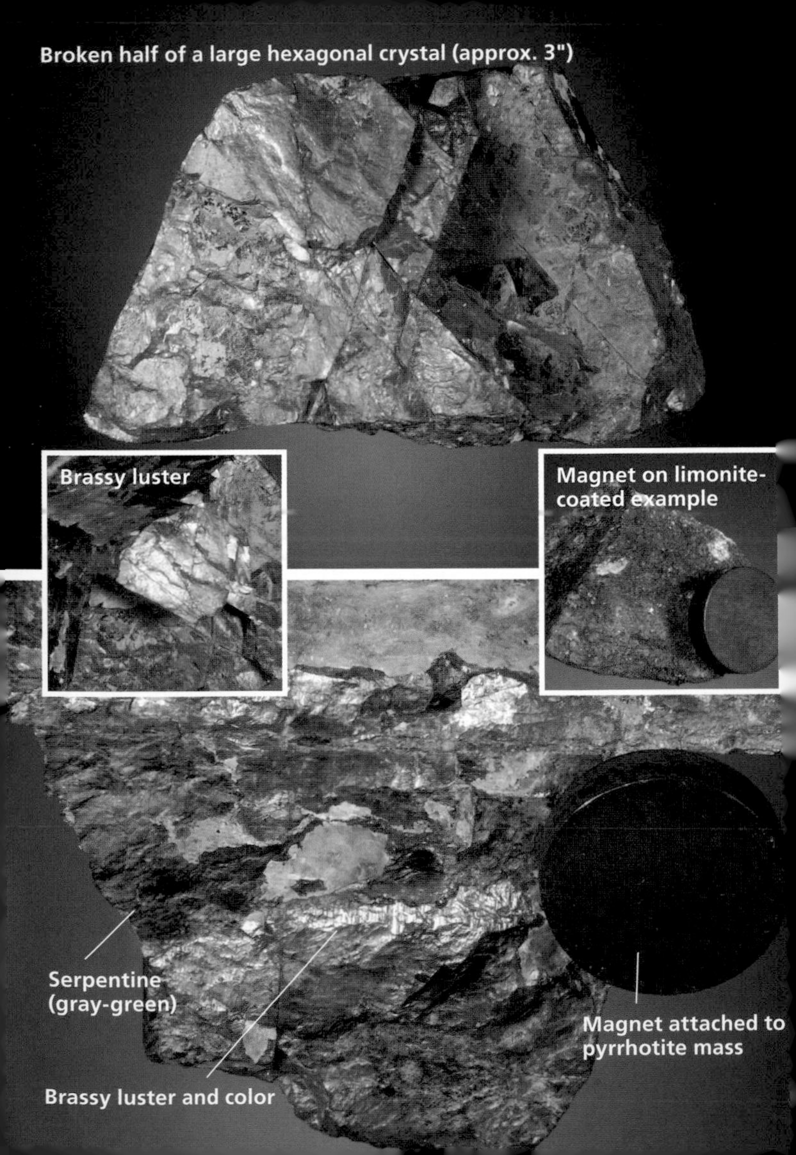
Broken half of a large hexagonal crystal (approx. 3")
Brassy luster
Magnet on limonite-coated example
Serpentine (gray-green)
Brassy luster and color
Magnet attached to pyrrhotite mass

Pyrrhotite

HARDNESS: 3.5-4.5 **STREAK:** Dark gray to black

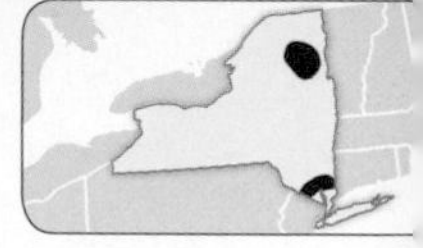

ENVIRONMENT: Mines, road cuts, quarries, outcrops

Primary Occurrence

WHAT TO LOOK FOR: Dense, metallic hexagonal crystals or masses with a brassy color and that are magnetic

SIZE: Specimens are typically no more than an inch or two but rarely can be fist-sized or larger

COLOR: Bronze-brown to brass-yellow, often brown on surfaces

OCCURRENCE: Rare

NOTES: Pyrrhotite is, like pyrite (page 177), a brassy, metallic mineral composed of iron sulfide, a combination of iron and sulfur. It differs from pyrite in a number of ways, however. The primary difference is its unstable molecular structure; unlike pyrite, pyrrhotite is missing iron atoms, and voids have taken their place. The iron deficiency creates an uneven chemical bond and makes pyrrhotite magnetic, which is a sufficient enough clue to distinguish it from the similarly colored minerals pyrite, marcasite (page 159), and chalcopyrite (page 75). Also distinctive is its crystal shape; though rarely present, pyrrhotite's crystals often appear intergrown as flat hexagonal (six-sided) plates. Masses or embedded grains are more abundant but still rare in the state; typically found in metamorphic rocks, such specimens can be identified by using a magnet, and this can help distinguish it from other brass-colored minerals. But bear in mind that pyrrhotite's characteristic luster and color can be hindered by a dull, dark brown tarnish that quickly develops on fresh surfaces.

WHERE TO LOOK: Exposures in the Adirondack Mountains have revealed small amounts of pyrrhotite, as have mines in Orange County, especially in the Tuxedo and Warwick areas. The best crystals came from Putnam County, however, where most were found in the Tiller Foster Iron Mine.

Large quartz crystal cluster
Pointed tip
Quartz in limestone
Double-terminated quartz crystal
Quartz (white) in granitoid
Crude quartz cluster
River-worn quartz

Quartz

HARDNESS: 7 **STREAK:** White

Primary Occurrence

ENVIRONMENT: All environments

WHAT TO LOOK FOR: Light-colored, translucent, glassy and very hard six-sided crystals; also found as masses, veins, and sand

SIZE: Crystals can be up to fist-sized; masses can be most any size

COLOR: Colorless to white or gray when pure, but often stained yellow to brown or red; more rarely black or pink

OCCURRENCE: Very common

NOTES: Quartz, the most abundant mineral in the earth's crust, is the first mineral all new collectors should study and be able to identify as it is ubiquitous around the world, including New York. It consists entirely of silica, the silicon- and oxygen-bearing compound that contributes to the formation of hundreds of other minerals, and it is most common as a component of rocks. As such, you'll encounter quartz most frequently as uninteresting white or gray grains or masses within coarse-grained rocks such as granite, or as the microscopic grains that make up chert. But crystals are not rare, either, and they most often take the form of elongated hexagonal (six-sided) prisms that end with a point, frequently intergrown in groups and lining cavities in rocks. Any river or beach will yield waterworn pebbles of quartz, and glacially deposited pebbles or fragments may be found in any gravelly area. Whatever its form when you find it, quartz is easy to identify due to its distinctively high hardness (it's typically the hardest mineral you'll find easily), its translucency, and its conchoidal fracture (when struck, the cracks that form are circular).

WHERE TO LOOK: Quartz can be found virtually anywhere, especially where there is loose gravel. It's a common find in rivers and on beaches because it's hard enough to survive weathering.

Rose quartz mass from pegmatite
Quartz druse
Smoky quartz crystals
Dolomite

Quartz, varieties

HARDNESS: 7 **STREAK:** White

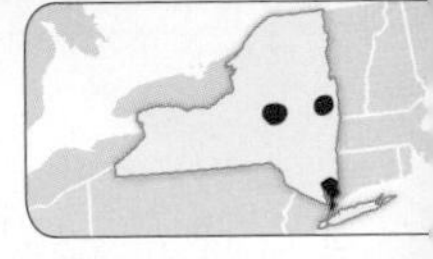

Primary Occurrence

ENVIRONMENT: All environments

WHAT TO LOOK FOR: Quartz crystals or masses with distinctly different colors and/or forms.

SIZE: Crystals can be up to several inches; masses can be virtually any size

COLOR: Gray to black, pink to purple, variegated or multicolored, depending on variety

OCCURRENCE: Common to rare, depending on variety

NOTES: Quartz forms in many different geological environments, each with its own set of variables. As a result, dozens of varieties of quartz exist in New York State and throughout the world. Quartz druse is common within pockets in limestone, and it consists of a crust of countless tiny crystals lining the inside of a cavity. Chert and jasper (page 143) are abundant microcrystalline forms of quartz, meaning that they are masses consisting of microscopic crystals or intergrown grains all tightly packed together, yet they still exhibit the traits of quartz. But some of the most popular varieties in the state are the colored quartzes such as smoky quartz (gray), rose quartz (pink), and amethyst (purple). Amethyst is rare in New York, but rose quartz is somewhat more common and found in pegmatites (very coarse-grained granite formations). Smoky quartz is especially popular in New York thanks to the prolific Herkimer region (see page 187), which produces brightly lustrous, nearly black crystals in limestone.

WHERE TO LOOK: The Herkimer area produces the fantastic water-clear "diamonds" of quartz as well as smoky quartz. Rose quartz is rare, found only in the pegmatites of Saratoga County, particularly near Greenfield.

Rough quartzite
Texture detail
Faint layering
River-worn examples

Quartzite

HARDNESS: ~7 **STREAK:** N/A

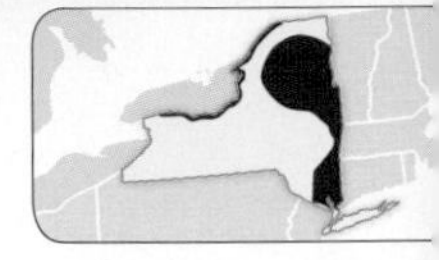

Primary Occurrence

ENVIRONMENT: Mountains, quarries, rivers, beaches, outcrops, hills

WHAT TO LOOK FOR: Very hard, tough, grainy rock with the general appearance and properties of quartz

SIZE: Quartzite can occur in any size, from pebbles to mountains

COLOR: Commonly white to gray, or brown, reddish, or yellow to green if impure; sometimes multicolored or banded

OCCURRENCE: Common

NOTES: Quartzite is an abundant metamorphic rock that forms when sandstone undergoes one of several processes. It typically forms either when sandstone is heated and compressed, causing the grains within the sandstone to fuse together, or when sandstone is saturated with silica (quartz-material), cementing the grains together. In either case, because the grains in sandstone are almost entirely quartz, the resulting quartzite is an extremely hard, quartz-rich rock that is highly resistant to weathering and white to gray when fairly pure but typically stained by iron or other minerals. Pebbles are common in gravel deposited by the glaciers or rivers, but whether found rounded and worn or freshly broken from an outcrop, it can be easy to confuse quartzite with quartz (page 185) itself. Quartz is generally always more translucent and appears glassier when broken, while quartzite has more of a grainy, flaky texture than quartz, particularly under magnification. Chert (page 77) may also seem similar to quartzite, but is always more opaque and is often waxy in appearance.

WHERE TO LOOK: Rivers and streams are great places to look. The Adirondacks are the best place to start because enormous portions of the mountains are composed of quartzite.

Sandstone
Texture detail
River-worn sandstone

Sandstone

HARDNESS: N/A **STREAK:** N/A

Primary Occurrence

ENVIRONMENT: All environments

WHAT TO LOOK FOR: Rocks with a very gritty texture and granular nature; you can often remove grains with your hands

SIZE: As a rock, sandstone can be found in any size

COLOR: Varies greatly; typically tan to yellow or brown, often reddish or orange, sometimes multicolored

OCCURRENCE: Common

NOTES: Many sedimentary rocks are named and defined by the grains from which they are composed; sandstone consists of sand, a type of sediment we're all familiar with. Sand is classified as detrital particles no larger than 1/12 inch and typically ends up being comprised of almost entirely quartz (page 185), due to its resistance to weathering. Sand turns to sandstone when it settles into beds at the bottoms of ancient seas and lakes, where pressure compresses it and other minerals, particularly calcite and clays, form between the sand grains, cementing them in place. Sandstone therefore consists largely of quartz, but as bonded grains, giving the rock a gritty, rough texture and making it possible to separate the grains fairly easily with your fingernails (though this depends on the level of cementation of a particular sandstone formation). Under low magnification (or sometimes none at all), the grains are easily visible, and layers of varying color, caused by iron or other minerals staining the cement, are common. Even highly weathered examples exhibit all of these traits, making sandstone perhaps the easiest sedimentary rock to identify.

WHERE TO LOOK: You'll find sandstone throughout most of New York; look in the Catskill Mountains, along the Mohawk River, or along any other rivers in the Finger Lakes region.

Mass of intergrown scapolite crystals

Fluorescing scapolite

Wernerite crystal on diopside

Massive gray wernerite in marble (shown fluorescing yellow-white under UV light on right)

Scapolite group

HARDNESS: 5-6 **STREAK:** White

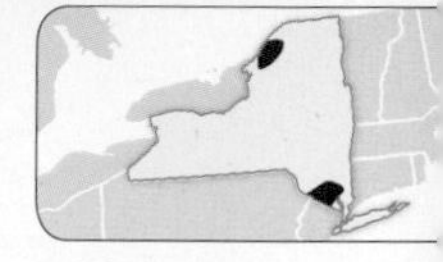

Primary Occurrence

ENVIRONMENT: Mines, quarries, outcrops, road cuts, hills, mountains

WHAT TO LOOK FOR: Light-colored, blocky crystals embedded in metamorphic rocks

SIZE: Crystals are generally an inch or two, but rarely larger

COLOR: White to gray, tan to yellow or brown, greenish

OCCURRENCE: Rare

NOTES: "Scapolite" is a name that no longer refers to a single mineral but rather to a family of aluminum- and silica-bearing minerals, the primary members being meionite, a calcium-rich scapolite, and marialite, a sodium-rich scapolite. Wernerite is technically a third member that contains both calcium and sodium, so it can be thought of as an intermediate mineral between meionite and marialite in composition. All are present in New York, though telling one from the other may be impossible outside a lab, especially when we consider that meionite often alters, or chemically changes, to become marialite. But distinguishing them as scapolite minerals is not quite as difficult. Formed primarily in metamorphic environments, scapolites appear as blocky, stubby, angular crystals, often dull in luster and uninterestingly colored, embedded in marble, skarn, or gneiss. They can greatly resemble feldspars (page 93) at times, but feldspars are more common and are slightly harder. In addition, scapolite minerals are frequently fluorescent under UV light; as feldspars rarely fluoresce, this can help you identify scapolites.

WHERE TO LOOK: The best localities are the Amity-Warwick area of Orange County and the Pierrepont area of St. Lawrence County, where scapolites can be found embedded in metamorphic rocks.

Mica schists
Pyrite crystal
Magnetite crystal
Texture detail
Brightly lustrous mica grains
Magnetite-rich schist with sillimanite crystals (tan)

Schist

HARDNESS: N/A **STREAK:** N/A

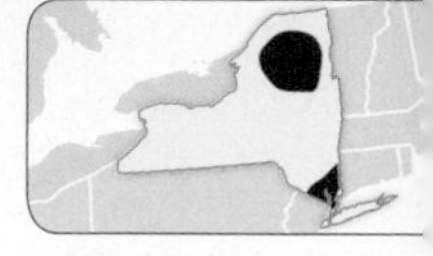

Primary Occurrence

ENVIRONMENT: All environments

WHAT TO LOOK FOR: Tightly layered, dense, fairly hard rocks often with copious mica and garnets

SIZE: As rocks, schists can be found in any size

COLOR: Varies; often multicolored in shades of white to gray or black, brown and green

OCCURRENCE: Common

NOTES: The rock gneiss (page 117) forms when rocks are heated and compressed, causing their minerals to break down and recombine into new minerals and organizing the resulting gneiss into layers. Schist takes this process even further. While gneisses tend to result from the metamorphism of granite or other crystalline rocks, schists can result from many rocks, particularly sedimentary rocks, and they undergo such an intense change that their original mineral grains are changed and recrystallized as completely different minerals. As a result, schists don't resemble the rocks from which they are derived. Instead, they appear tightly layered and compact, often fairly hard, and frequently with large amounts of micas, often in such quantity that the schist appears "glittery." Other hard minerals, such as garnets, often form in schists, which makes them an attractive rock for collectors. Schists are named for their predominant mineral, such as mica schists, which consist primarily of micas; others are rich with talc, serpentines, magnetite, and many other minerals.

WHERE TO LOOK: Schists are common in the Adirondacks and the surrounding counties, as well as in the far southeastern corner of the state, near New York City. They are also frequently found in gravel and on shores throughout the state, due to glacial activity.

karn containing blue masses of serendibite

Blue grains of serendibite in skarn

Serendibite

HARDNESS: 6.5–7 **STREAK:** White

Primary Occurrence

ENVIRONMENT: Mines

WHAT TO LOOK FOR: Hard, blue granular masses or prisms

SIZE: Crystals are tiny, while masses may be up to several inches

COLOR: Light to dark blue, greenish blue

OCCURRENCE: Very rare

NOTES: Serendibite is a very rare boron-bearing mineral; it is found in only a handful of places worldwide. Nonetheless, serendibite was first discovered in New York, and there are multiple locations in New York where it can be found, making it seem that New York is unusually rich with this hard-to-find mineral. In reality, while it did form in several regions throughout the state, it remains extremely scarce. Serendibite forms in metamorphic rocks, particularly marble and skarn, through a process called contact metamorphism in which a body of molten rock, often granite, contacts preexisting rock, which in this case was limestone or dolostone. Contact metamorphism is not uncommon, but a source of boron is not always present under such conditions. Serendibite crystals are blocky or tabular (flattened), but massive or granular examples are far more common in New York; these are found embedded in their host rock where serendibite is most identifiable by its characteristic blue color, glassy luster and high hardness. These traits, in combination with its rare occurrence, make serendibite fairly conspicuous in New York State.

WHERE TO LOOK: Few mines in New York ever produced this rare mineral; mines in the Amity, Warwick, and Edenville areas of Orange County and near Johnsburg in Warren County are the most prominent localities. Unfortunately, these areas are off-limits to collectors unless permission is obtained from landowners.

Grooved serpentine mass
"Greasy" luster
Reddish serpentine
Fibrous chrysotile vein
Lustrous serpentine mass

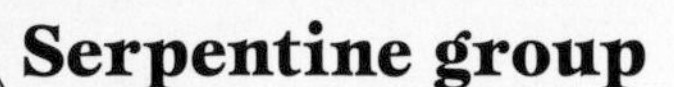

Serpentine group

HARDNESS: 2.5-5 **STREAK:** White to greenish white

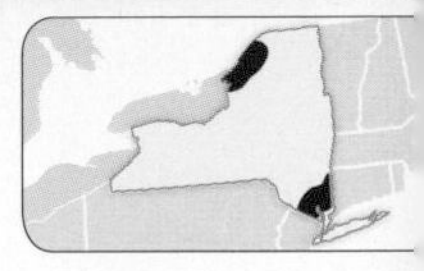

Primary Occurrence

ENVIRONMENT: Mines, hills, outcrops, road cuts

WHAT TO LOOK FOR: Fairly soft, green, silky or "greasy" masses, often striated or grooved and found in metamorphic rocks

SIZE: Masses of serpentine can range greatly, up to boulder-size

COLOR: Light to dark green common, also tan to brown, gray to black, or rarely reddish or purple; often mottled

OCCURRENCE: Uncommon

NOTES: The serpentines are a group of closely related minerals with some unique properties and varieties that make them fun to collect and often easy to identify. They generally develop within, or as a constituent of, metamorphic rocks, and form when magnesium-bearing minerals, such as olivines, are altered. This means that New York's most abundant serpentines—antigorite, lizardite, and chrysotile—are found closely associated with rocks like schist and marble. Generally opaque and in hues of green or yellow, the serpentines are all fairly soft and easily scratched with a knife (unless they are incorporated in a harder rock, in which case serpentines are often found in layers). Particularly pure specimens often have a distinctly "greasy" or slippery feel to them. These traits that make it quite easy to differentiate serpentines from other green minerals make it hard to tell them apart from each other. The exception is chrysotile, which typically consists of tightly compact, flexible, thread-like fibers with a silky sheen; this is a variety of asbestos, so care should be taken when working with it. Several serpentine minerals can also occur together; such rocks are referred to as serpentinite.

WHERE TO LOOK: Putnam, Warren, and other eastern counties have produced serpentines, but no locality has been more prolific than the Tilly Foster Iron Mine near Brewster.

Sheets of layered shale
Easily separated layers
Fossil shell in shale
Rough flaky shale
River-worn shale

Shale

HARDNESS: <5.5 **STREAK:** N/A

Primary Occurrence

ENVIRONMENT: Lowlands, hills, quarries, rivers, outcrops, road cuts, beaches

WHAT TO LOOK FOR: Soft, fine-grained rocks in flat, layered sheet-like formations; layers can be separated with a knife

SIZE: As a rock, shale can be found in any size

COLOR: Tan to yellow or brown, light to dark gray or black

OCCURRENCE: Very common

NOTES: Throughout its geological history, New York was repeatedly submerged beneath shallow seas; this caused a variety of sedimentary rocks to form. Shale, one of the most common in the state, is a highly layered rock that forms when beds of mud are compacted and solidify at the bottoms of very still bodies of water. It consists of microscopic grains of other rocks and minerals that have weathered, particularly micas, clays, and some quartz; these were deposited periodically, giving shale its layered habit. But as a result, shale is a soft rock and easy to scratch with a knife; its layers can easily be separated and pulled apart, sometimes revealing fossils hidden between. In addition, soaking a piece in water can soften it further; all of these traits are distinctive and make shale easy to identify. Mudstone (page 169) is extremely similar and shares most traits with shale, except one: the layering. Mudstone is, in simplest terms, unlayered shale that formed in more turbulent waters. Most of New York's fossils, including the famous eurypterids (page 103), are found in shale, and virtually any outcrop could yield traces of ancient life.

WHERE TO LOOK: Like limestone and sandstone, shale is pr in much of New York, particularly in the southern a central areas. Virtually any river, beach, or outcrop Finger Lakes area will reveal layered samples.

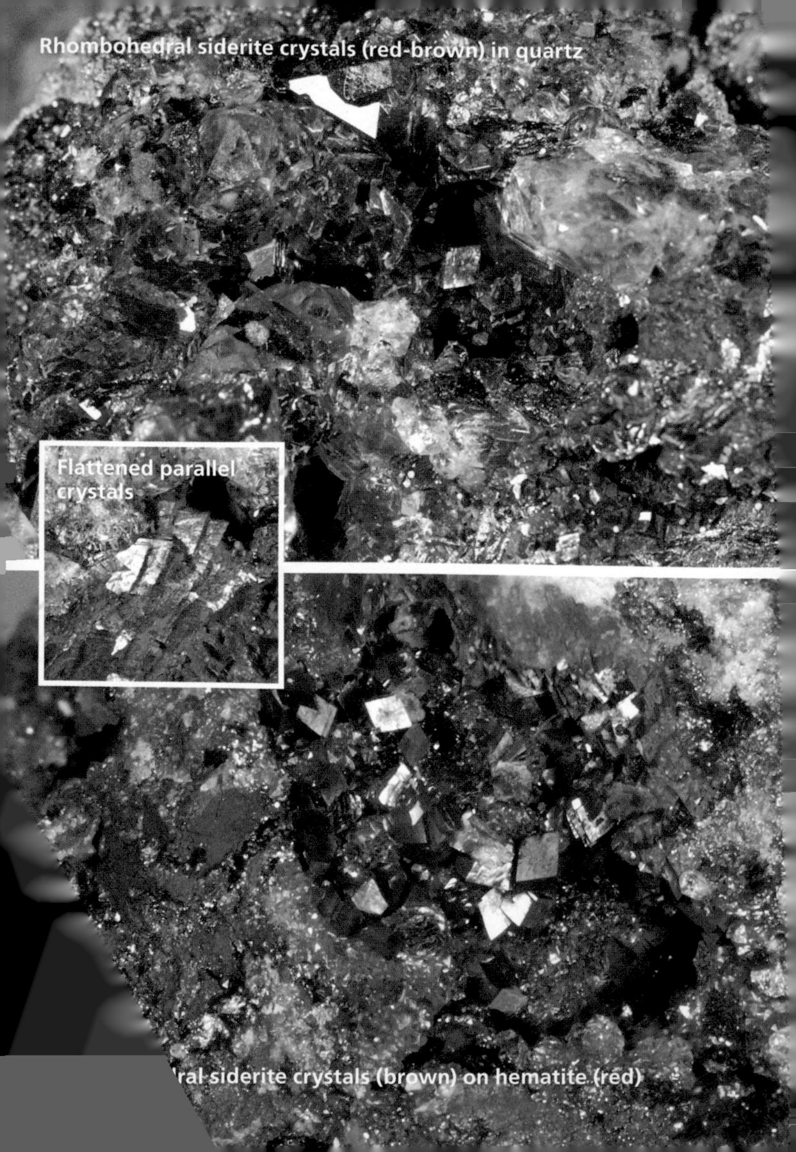

Rhombohedral siderite crystals (red-brown) in quartz

Flattened parallel crystals

ral siderite crystals (brown) on hematite (red)

Siderite

HARDNESS: 3.5-4 **STREAK:** White

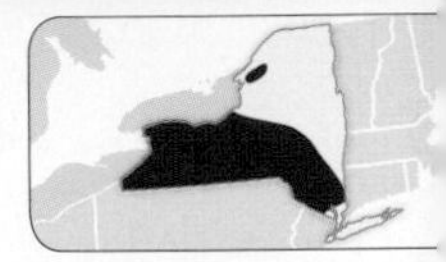

Primary Occurrence

ENVIRONMENT: Mines, quarries, outcrops, road cuts, hills, mountains, lowlands

WHAT TO LOOK FOR: Small, brownish, pearly, rhombohedral crystals or masses with step-like surfaces

SIZE: Crystals remain smaller than your thumbnail, but masses may be up to several inches

COLOR: Tan to brown or dark brown, rusty red, yellow

OCCURRENCE: Uncommon

NOTES: Composed of iron carbonate, siderite is a close mineral relative to calcite (page 63), and to dolomite (page 87), with which it is frequently confused. All three minerals share the same primary crystal shape of a rhombohedron, which resembles a leaning or skewed cube, though siderite's crystals are often seen more flattened in shape and resemble a disk or blade, often with curved edges. In virtually all cases, siderite specimens are opaque and light-colored with a pearly luster, and they are generally found alongside other iron-bearing minerals such as pyrite or hematite. But massive specimens are more common and often have a very indistinct appearance, making them easy to overlook. In these cases, siderite's perfect rhombohedral cleavage (when struck, it breaks in rhombohedral shapes and angles) is distinctive, and specimens often have steeply stepped surfaces. Calcite tends to be more translucent than siderite, as does dolomite; calcite is also softer than siderite, but dolomite is too similar for hardness to help. Instead, siderite will become slightly magnetic when heated in a flame; dolomite will not.

WHERE TO LOOK: The mines in Antwerp in Jefferson County produce the best specimens in the state, but much of the area is private or protected.

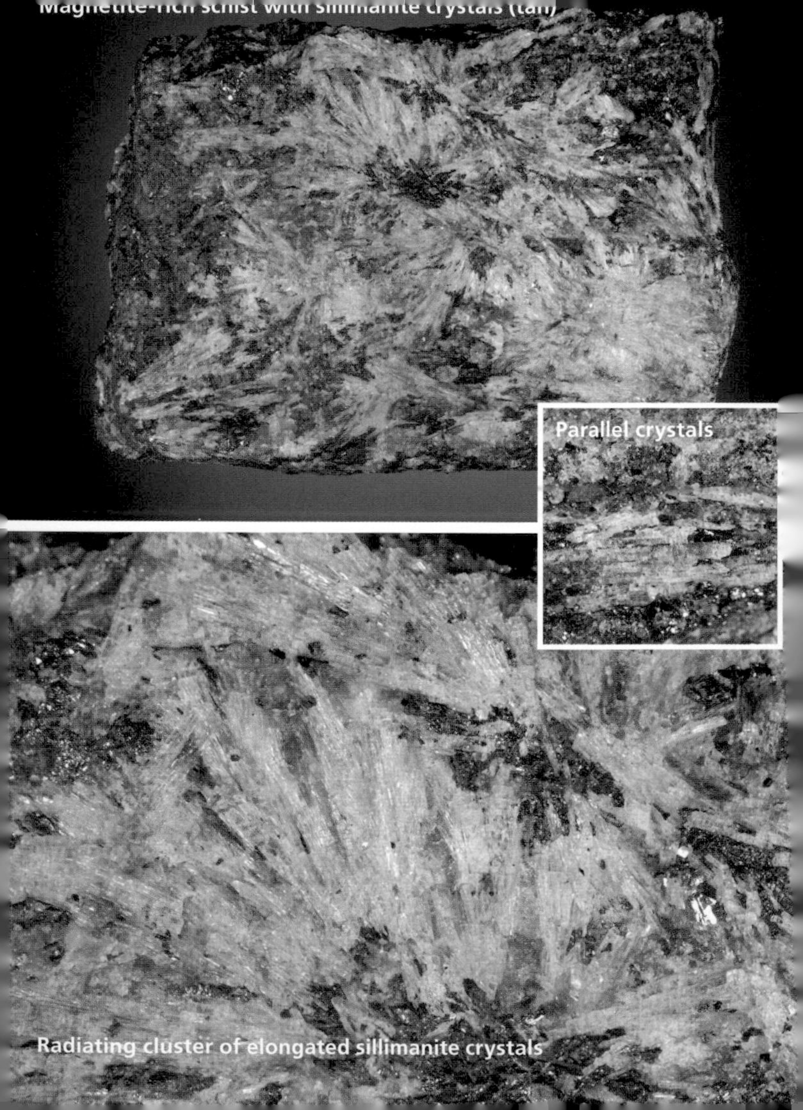
Parallel crystals
Radiating cluster of elongated sillimanite crystals

Sillimanite

HARDNESS: 6.5–7.5 **STREAK:** White

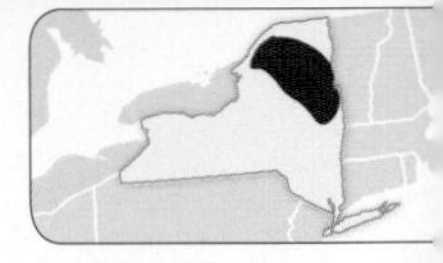

Primary Occurrence

ENVIRONMENT: Mines, outcrops, road cuts, mountains

WHAT TO LOOK FOR: Elongated, fibrous, generally rectangular crystals embedded in metamorphic rocks, particularly gneiss

SIZE: Rare crystals can measure six inches in length, but most are smaller than an inch

COLOR: White to gray or black, light to dark brown, red-brown

OCCURRENCE: Uncommon

NOTES: Sillimanite forms in metamorphic rocks, such as gneiss and schist, as they undergo the heat and pressure of metamorphic activity. Sillimanite consists of aluminum and silica, and like garnets, it developed as the minerals present in metamorphic rocks changed, broke down, and recombined into different minerals during metamorphosis. In New York, sillimanite is found embedded in its host rocks, gneiss and schist, often alongside feldspars and magnetite. Sillimanite crystals occur as elongated, vaguely rectangular prisms with a distinctly fibrous or silky appearance. Sometimes the crystals may be arranged in radial patterns as well. Sillimanite is often gray or tan in color and always very hard, but it is brittle and splinters easily when struck. Many of its traits make it easy to confuse with fibrous members of the amphibole group (page 39), though it is typically harder and rarer. Amphiboles also exhibit well-defined crystal shapes more often, while sillimanite typically isn't so distinct, with many specimens appearing as fibrous "streaks" in their host rocks.

WHERE TO LOOK: Essex, Clinton, Saratoga, and Warren Counties have produced specimens in the metamorphic rock outcrops of the Adirondacks. Some of the best specimens, however, originated from St. Lawrence County in the Benson-Fine area.

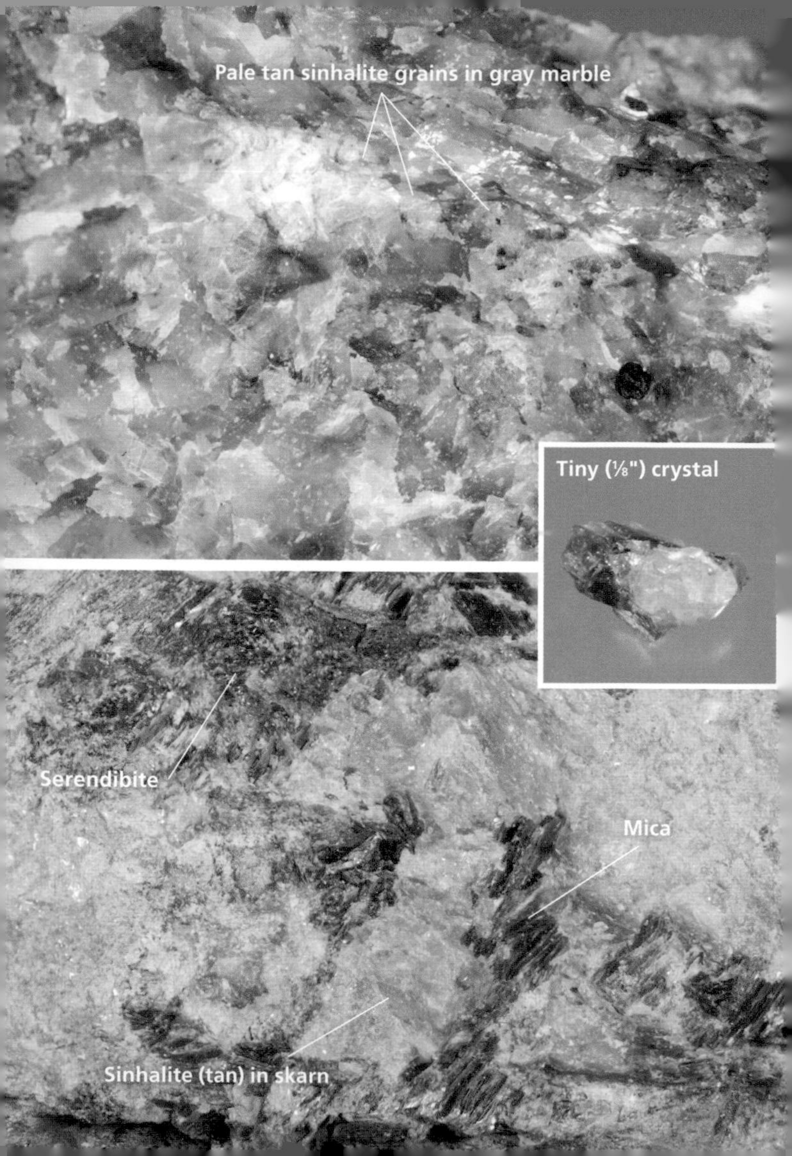
Pale tan sinhalite grains in gray marble
Tiny (⅛") crystal
Serendibite
Mica
Sinhalite (tan) in skarn

Sinhalite

HARDNESS: 6.5–7 **STREAK:** Colorless

Primary Occurrence

ENVIRONMENT: Mines

WHAT TO LOOK FOR: Inconspicuous, very hard, light-colored grains or masses in marble or skarn

SIZE: Specimens are generally just grains, smaller than a pea

COLOR: Tan to pale yellow, sometimes pink or flesh-colored

OCCURRENCE: Very rare

NOTES: Sinhalite was first found in Sri Lanka and gets its name from the Sanskrit word for Sri Lanka. Worldwide, sinhalite is extremely rare, but it is found in more than one place in New York thanks to the state's uncommon metamorphic geology. Found only in marble or skarn that is unusually enriched with boron, sinhalite is almost never found crystallized and is instead seen only as tiny glassy grains or masses. Since it is always light-colored and generally tan, it can be very inconspicuous when found in host rock of a similar color, making specimens easily and frequently overlooked. This also makes sinhalite difficult to identify, and one must use the surrounding rocks and minerals as a guide. If a specimen is large enough, it can be tested for hardness as sinhalite is very hard and few other minerals found in its environment will match it, but otherwise you can look for associated minerals such as the equally rare serendibite (page 197) as well as spinel (page 215) and phlogopite mica (page 161). Though it isn't a particularly attractive mineral, sinhalite's rarity makes it a compelling find and specimens are often valuable.

WHERE TO LOOK: The most significant specimens originated fro the Johnsburg Township area of Warren County and fror the Edenville area mines of Orange County. Both areas, however, are largely off-limits to collectors unless perr is obtained, so buying a specimen may be your only

Serendibite (blue)
Wollastonite (white
Rough specimens of skarn
Coarse scapolite crystals in skarn
Skarn
arnet
Wollastonite

Skarn

HARDNESS: N/A **STREAK:** N/A

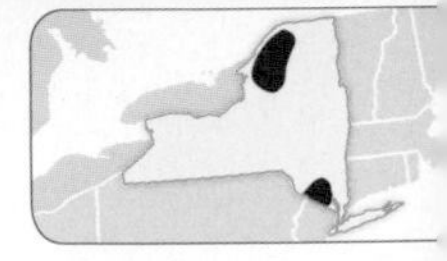

Primary Occurrence

ENVIRONMENT: Hills, mountains, mines, quarries, outcrops, road cuts

WHAT TO LOOK FOR: Coarse-grained rock associated with marble that contains many different minerals, often with open pockets containing finer crystals

SIZE: As a rock, skarn can be found in any size

COLOR: Varies greatly; mottled and multicolored, with white, tan to brown, yellow to green, reddish

OCCURRENCE: Uncommon

NOTES: Skarn is a metamorphic rock formed by what is called contact metamorphism, a process in which magma (molten rock) contacts preexisting rock, in this case limestone, heating the rock and exposing it to hot, mineral-rich water. This alters the limestone and enriches it with minerals; as it cools, large crystals develop, resulting in skarn, a coarsely crystalline rock of varying composition. Skarn is frequently confused with marble, due to their similar metamorphic origins, but the two aren't difficult to distinguish. Marble tends to be solid and compact with few voids, whereas skarn frequently contains cavities in which well-formed crystals can be found. Skarn's texture can also be variable, with finer-grained portions surrounding coarser ones. Green pyroxenes, black amphiboles, brown garnets, and white scapolites are common sights in skarn, and as many of these minerals can be well-formed and therefore highly collectible, the rock is a popular source of specimens.

WHERE TO LOOK: The Amity-Warwick area of Orange County as well as multiple localities in St. Lawrence County, are key skarn locations and produce fine specimens.

Broken masses of slag
Colored layers

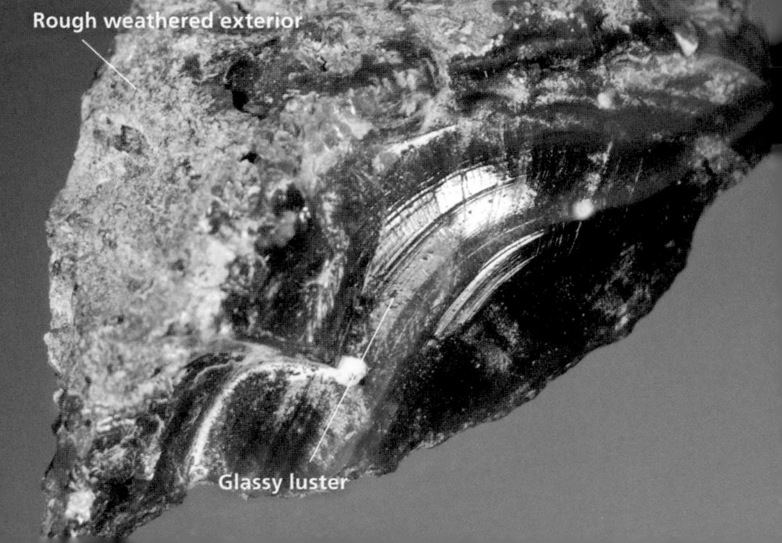
Rough weathered exterior
Glassy luster

Slag

HARDNESS: N/A **STREAK:** N/A

Primary Occurrence

ENVIRONMENT: Mines, quarries, rivers

WHAT TO LOOK FOR: Heavy, glassy to metallic masses with no particular shape, often with swirls of color or bubbles

SIZE: Slag may be any size, but typically smaller than a fist

COLOR: Varies greatly; typically dark-colored in shades of brown, green, blue to purple, or black

OCCURRENCE: Uncommon

NOTES: Many industrial processes produce unwanted by-products such as slag. Smelting is one such process; it involves melting ore minerals at certain temperatures to release only the desired metals. The rest of the molten material is discarded and rapidly cools and solidifies. The result is slag, a hard, often dense, glassy or metallic looking solid with no distinct shape of its own. Slag is essentially a "blend" of destroyed minerals and a specimen's look can take on many appearances, including everything from wavy to rope-like textures, colored layers, or containing countless bubbles. A specimen's look depends on how and where it cooled, and many variations are possible. In general, all slag is quite dark-colored, often in shades of green to bluish or black, and glass-like slag can be translucent while more metallic slag is typically opaque. It usually has conchoidal fracture (when struck, it breaks in a circular pattern), which can make it easy to mistake for jasper (page 143) or other quartz minerals. Slag, however, is often too smoothly textured to be like most minerals, is often very dense, and can contain trapped gas bubbles, which very rarely occurs in minerals.

WHERE TO LOOK: Any district that was once home to mining or other forms of industry may yield slag. In addition, slag has been used along railroads as fill material, though be aware that railroads are private property.

Very fine translucent sphalerite (red-brown) on dolomite crystals

Yellow crystal cluster

Complex crystal

Cleavage planes

Large pure mass of sphalerite

Sphalerite

HARDNESS: 3.5-4 **STREAK:** Pale yellow to brown

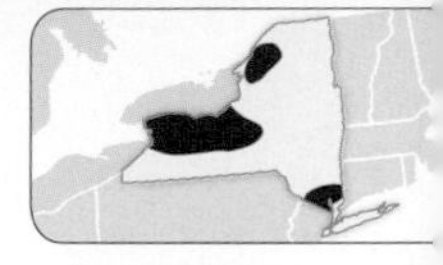

Primary Occurrence

ENVIRONMENT: Mines, quarries, outcrops, road cuts, hills, lowlands, mountains

WHAT TO LOOK FOR: Soft, lustrous, often vividly colored crystals with complex angular shapes and triangular faces

SIZE: Crystals are usually under an inch; masses may be fist-sized

COLOR: Yellow to brown, red, black; less frequently green

OCCURRENCE: Common

NOTES: As the primary ore of zinc, sphalerite is an economically important mineral and collectors cherish it for its fantastic crystal specimens. A zinc sulfide, it forms very commonly alongside chalcopyrite (page 75) and other sulfide minerals, and it is commonly found within sedimentary rocks such as limestone, but it's also found within ore veins. Sphalerite crystals are not rare and can be very complexly shaped due to their common habit of twinning (when two or more crystals grow within each other). Crystals typically have triangular faces and wedge-shaped points. Sphalerite's color can vary greatly, from golden yellow to deep red or even black, but it is always brightly lustrous unless heavily weathered, and is typically translucent. Very dark, impure specimens can be opaque and almost metallic in appearance, making them possible to confuse with galena (page 113), with which it can occur, though galena is much softer. The easiest specimens to identify are those from limestone deposits, where crystals often form in cavities on top of dolomite crystals (page 87).

WHERE TO LOOK: The Balmat-Edwards area mines in St. Lawrence County primarily produced sphalerite, including fine crystals, as well as enormous masses, but the area is largely inaccessible today. Wayne and Niagara County quarries also produce spectacular crystals on dolomite; these can still be collected.

Spinel crystals freed from marble

Bluish crystal

Magnetite crystal

Octahedral spinel crystal

Spinel group

HARDNESS: 5.5-8 **STREAK:** Gray to black

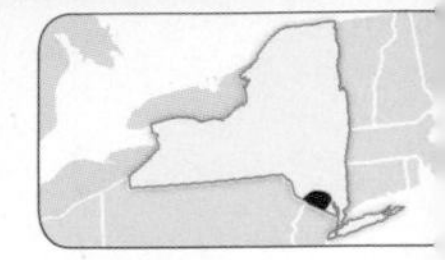

Primary Occurrence

ENVIRONMENT: Mines, outcrops, road cuts, mountains, hills, quarries

WHAT TO LOOK FOR: Hard, eight-faced, sharply pointed crystals with pyramid-like shapes embedded in rock

SIZE: Different group members may be found in a range of sizes, but most are generally smaller than an inch

COLOR: Dependent on the group member; black to gray or brown is common, also dark green to dark blue

OCCURRENCE: Uncommon to very rare, depending on the type

NOTES: The spinel group is a family of oxide minerals, with each group member being composed of two metallic elements combined with oxygen. All spinel minerals share the same characteristic octahedral crystal shape—an octahedron is an eight-faced crystal resembling two four-sided pyramids placed base-to-base. Crystals, when present, are often well-formed with sharply defined edges and points. A few group members are found in New York, but magnetite, an iron oxide, and spinel, a magnesium-aluminum oxide and primary group member, are the most prominent. Spinel is very hard—not even quartz (page 185) will scratch it—and it exhibits a glassy luster and is colored in dark shades of gray or occasionally green. While rare, it is most frequently found in the marble in the southeastern corner of the state, where it is conspicuous as dark, hard grains in the softer surrounding rock. Magnetite, discussed in detail on page 155, is considerably different, as it is metallic, much softer, found in more environments, and, most notably, magnetic.

WHERE TO LOOK: Most localities are centered around the Warwick-Amity and Monroe areas in Orange County. Specimens are found embedded in marble in this historic mining area.

Cluster of white strontianite needle-like crystals
Baryte
Crystal cluster

Strontianite

HARDNESS: 3.5 **STREAK:** White

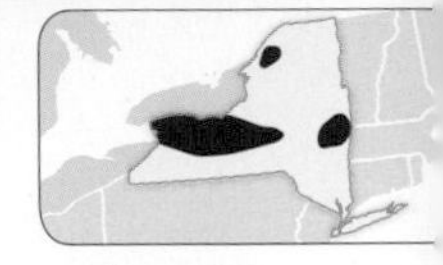

Primary Occurrence

ENVIRONMENT: Mines, quarries, lowlands, hills

WHAT TO LOOK FOR: Small, white, needle-like crystals arranged into ball-like groups, often with calcite or baryte

SIZE: Most crystals are tiny, smaller than ⅛ inch

COLOR: Colorless to white, gray to light brown

OCCURRENCE: Rare

NOTES: Strontianite is a scarce mineral that is easily overlooked or confused with other more common, similar-looking minerals. Closely related to aragonite (page 49), strontianite is composed of strontium carbonate and forms primarily within cavities in sedimentary rocks, particularly limestone (page 149), where it appears most often as tiny, delicate, needle-like crystals. These crystals can be singular, but are more commonly grouped together and arranged into radial spray- or ball-like groupings. Strontianite is generally white in color and glassy, and since its crystal shape is often too small to easily observe, these traits make it easy to confuse with calcite (page 63), aragonite, and other more abundant minerals. Hardness is difficult to test on tiny specimens, so with strontianite you may be left guessing. However, there is one other trait that may help: fluorescence. Strontianite is so often fluorescent under UV light that it can be an identifying trait; it glows in shades of white, pale yellow, or a faint blue. Calcite and other similar minerals may fluoresce and in similar colors, but they do less frequently.

WHERE TO LOOK: Widespread but in very small amounts, Niagara, Albany, and St. Lawrence County mines and quarries have produced specimens, particularly within limestone, but generally only as tiny crystals.

Polished moonstone specimen showing bright white internal schiller

"Sunstone"/"Moonstone"

HARDNESS: 6-6.5 **STREAK:** White

Primary Occurrence

ENVIRONMENT: Mountains, mines, quarries, outcrops, road cuts, rivers

WHAT TO LOOK FOR: Hard, blocky masses exhibiting internal colorful reflections or schiller of orange or white to blue

SIZE: Both occur massively and can be found up to a foot in size

COLOR: Sunstone is tan to orange with colorful internal spots; moonstone is white to gray with bluish internal flashes

OCCURRENCE: Sunstone is very rare; moonstone is rare

NOTES: Similar not only in name but also in composition, sunstone and moonstone are both rather unique varieties of feldspar (page 93) and formed in coarse granitic rocks. Sunstone, which is typically microcline or oligoclase feldspar, is generally orange in color and has tiny reflective spots within its translucent structure that reflect light, often in a rainbow of colors. These spots are actually tiny hematite (page 129) inclusions within the feldspar. Moonstone, on the other hand, is also typically microcline feldspar, but exhibits a bright white-blue internal schiller (flashes of color) as it is rotated under a bright light. This trait is derived from the layered internal structure of a feldspar mass, which actually consists of many parallel crystals, and causes the light to bounce between them, much like the similar schiller phenomenon seen in labradorite (page 145). Both are identified using the typical traits of feldspar along with the unique and highly desirable visual characteristics described above.

WHERE TO LOOK: Sunstone is rarer in New York but has been found in the Benson Mines, near Fine in St. Lawrence County as well as in the Chappaqua area in eastern New York. Moonstone is more common and is frequently collected in the rocks of the Adirondacks around Saranac Lake.

Masses of talc containing needles of tremolite
Green crystal cluster
Mass of talc

Talc/Soapstone

HARDNESS: 1 **STREAK:** White

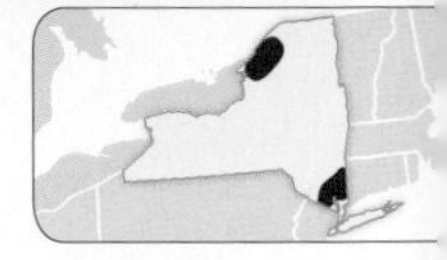

Primary Occurrence

ENVIRONMENT: Mines, outcrops, road cuts, mountains, hills, quarries

WHAT TO LOOK FOR: Extremely soft light-colored masses, often with a distinctly "soapy" feel and/or fibrous appearance

SIZE: Masses of talc and soapstone can be any size, and frequently measure several feet

COLOR: White to gray, light to dark green, yellow to brown

OCCURRENCE: Uncommon

NOTES: At a hardness of just 1 on the Mohs scale, talc is the softest mineral you'll encounter in New York, making it also one of the easiest to identify. Simple mishandling will scratch a specimen, and this extremely low hardness also lends talc its frequently "soapy" or "slippery" feel. Combined with its typically light coloration, often white to pale green, there are few things talc can be confused with. Talc forms when magnesium-rich minerals such as serpentine (page 199) undergo metamorphosis, and it is therefore most often found in metamorphic rocks such as schist, often alongside the minerals from which it was derived. Because of its metamorphic past, crystals are very rare and talc is far more abundant as irregular, layered masses that occasionally have a fibrous appearance. Masses of talc are rarely pure, however, and other minerals, particularly amphiboles (page 39), can be intergrown within the talc, making a flaky, layered, talc-rich rock called soapstone, named for its characteristically "soapy" texture. In New York, talc's association with serpentines is a key clue to its identity.

WHERE TO LOOK: Talcville, in St. Lawrence County, is home to many specimen-producing mines. Putnam County iron mines produced nice specimens, still obtainable today.

Tetrahedrite crystal with colorful reflections under bright light

Tetrahedrite group

HARDNESS: 3.5-4 **STREAK:** Black to brown

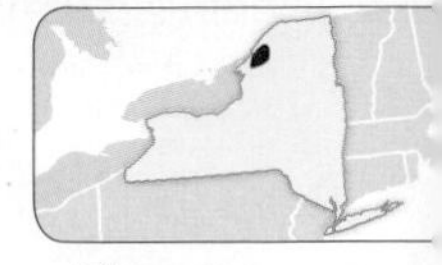

Primary Occurrence

ENVIRONMENT: Mines

WHAT TO LOOK FOR: Tiny, dark, tetrahedral (four-faced) metallic crystals, often with sphalerite

SIZE: Crystals are rarely larger than ½ inch, usually much smaller

COLOR: Steel-gray; may have colorful surfaces in bright light

OCCURRENCE: Very rare

NOTES: There are three primary minerals in the tetrahedrite group—tennantite, freibergite, and tetrahedrite itself—all of which are present in New York, but in very small amounts. They are so similar that specimens are typically just labelled "tetrahedrite," since they all share nearly identical traits. Tetrahedrites are complex sulfosalt minerals (a combination of a metallic element and a semi-metallic element with sulfur) and develop as tetrahedral crystals, which are triangular pyramids, often with interesting modifications, or variations of the crystals' edges or faces. This crystal shape is so distinctive that it is often enough for identification, but less ideal crystals or massive specimens are more difficult to identify, as its hardness and streak are not especially unique; nonetheless, its usual association with sphalerite (page 213), galena (page 113), or pyrite (page 177) will help. In addition, crystallized surfaces often show subtle triangular markings, which indicate its crystal structure, and under bright light specimens may also show bluish or multicolored hues.

WHERE TO LOOK: Tetrahedrite group minerals were rarely found in the Balmat-Edwards area zinc mines in St. Lawrence County, but specimens are difficult to come by. Collecting your own specimen is difficult today but may still be possible with persistence. Specimens do come up for sale, albeit rarely.

Fine cluster of parallel dravite crystals

Dravite cluster

Dravite in calcite

Fluor-uvite crystal

Tourmaline group

HARDNESS: 7-7.5 **STREAK:** White

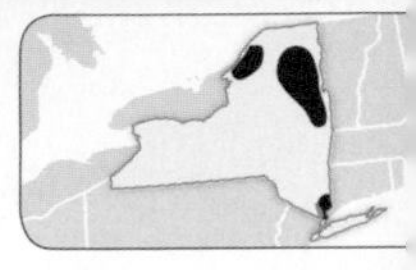

Primary Occurrence

ENVIRONMENT: Hills, mountains, mines, quarries, road cuts, outcrops

WHAT TO LOOK FOR: Very hard, elongated, lustrous blocky crystals, sometimes with striated (grooved) faces, typically dark in color, and embedded in quartz or other minerals

SIZE: Crystals can be up to several inches in length, while clusters can be up to a foot

COLOR: Commonly black; also gray to brown, rarely green

OCCURRENCE: Rare

NOTES: The tourmalines are a family of chemically complex minerals that form both in coarse igneous rocks, like pegmatite, as well as in metamorphic rocks, like skarn. They are some of New York's most famous minerals. Dravite, uvite, and schorl are the primary group members found in the state, and all of them are very hard and form as elongated, angular prisms, typically with bright, glassy luster and sometimes with deeply striated (grooved) faces. The state's dravite crystals, particularly those from the Pierrepont area, are legendary and among the finest in the world, appearing as clusters of opaque, glossy black, blocky crystals alongside quartz and amphiboles. Uvite and its fluorine-rich variety, fluor-uvite, are also famous in the state, appearing as stubby brown, translucent crystals with tremolite in metamorphic areas. Lastly, schorl appears as elongated, striated prisms, often within quartz and is found in granite or pegmatite outcrops. The tourmalines' high hardness and luster are distinctive, which distinguishes them from similar amphiboles (page 39).

WHERE TO LOOK: The Pierrepont and De Kalb areas of St. Lawrence County are world-famous for fantastic lustrous dravite. Outcrops in the Adirondacks have also produced tourmalines.

Masses of tightly intergrown tremolite crystals

Crystal mass detail

Gemmy green tremolite

Tan glassy tremolite mass (shown fluorescing blue and green under UV light on left)

Tremolite

HARDNESS: 5-6 **STREAK:** White

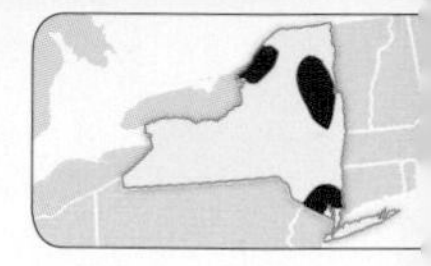

Primary Occurrence

ENVIRONMENT: Mines, quarries, outcrops, road cuts, hills, mountains

WHAT TO LOOK FOR: Elongated, fibrous, hard, light-colored crystals or masses in metamorphic rocks

SIZE: Crystals can be up to an inch or two; masses are up to a foot

COLOR: White to gray if pure; brown, pale to dark green if impure

OCCURRENCE: Uncommon

NOTES: Tremolite is one of the most prominent amphibole group (page 39) minerals in New York State, with dozens of localities producing collectible examples. Tremolite generally forms as elongated, fibrous crystals or masses within metamorphic rocks such as marble, and it is generally very light in color. Specimens containing iron are an exception, and these develop a deeper green color as the iron content increases. Once the iron content is above a certain threshold, the specimen technically becomes actinolite (page 37), a very closely related mineral, but differentiating between green tremolite or actinolite requires a laboratory. Tremolite's splintery, fibrous texture and often silky sheen is distinctive, but it is also fairly hard, which will help distinguish it from other similar fibrous minerals such as chrysotile (page 199). Some specimens are so fibrous they seem nearly fabric-like and are referred to as "mountain leather" (page 167); these are a variety of asbestos (fibrous minerals cancerous if inhaled), so a respirator is recommended when handling. Lastly, some specimens are fluorescent in UV light.

WHERE TO LOOK: Localities in St. Lawrence County have continually produced some of the world's finest specimens. The Balmat-Fowler and Gouverneur areas are the primary mining districts where specimens were produced.

Fine cluster of hexagonite crystals
Hexagonite crystal
Chrome-tremolite crystals
White tremolite
Chrome-tremolite (green)

Tremolite, varieties

HARDNESS: 5-6 **STREAK:** White

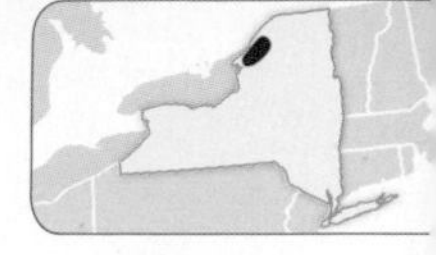

Primary Occurrence

ENVIRONMENT: Mines

WHAT TO LOOK FOR: Mineral formations with the traits of tremolite, but with a vivid coloration

SIZE: Crystals can be up to an inch or two; masses and clusters can be up to a foot or more

COLOR: Pink to purple; vivid emerald green to mint green

OCCURRENCE: Very rare

NOTES: Collectors searching the internet or rock shops for New York tremolite specimens will inevitably come upon the state's most spectacular examples: a pink to purple variety known as "hexagonite" and a vivid green variety called "chrome-tremolite." Hexagonite is exclusive to New York, and is found nowhere else in the world, and chrome-tremolite is incredibly rare as well—as a result, both are rare, popular, and valuable. Neither are distinct minerals, but are merely colored varieties that result from impurities that change tremolite's color. Hexagonite's lilac-hues are caused by manganese, and chrome-tremolite's rich emerald-green color is caused by chromium, as its name implies. Identification by color alone is easy, but looking for tremolite's typical traits—fibrous, fairly hard, elongated crystals with silky sheen and translucency—will eliminate any doubt. Unfortunately, collecting these rarities today is next to impossible, as the mines that produced them yielded small amounts and have long been shut down, but specimens are still found on the marketplace.

WHERE TO LOOK: Hexagonite is most famous from the Balmat-Fowler and Gouverneur area mines in St. Lawrence County, but those mines are long closed. The rare chrome-tremolite was only produced at the American Talc Mine in Balmat, now also closed.

Rocky mass containing tiny gray grains of turneaureite (virtually indistinguishable in normal light)

Same specimen as above under shortwave UV light (turneaureite uorescing vivid orange)

Turneaureite

HARDNESS: 5 **STREAK:** White

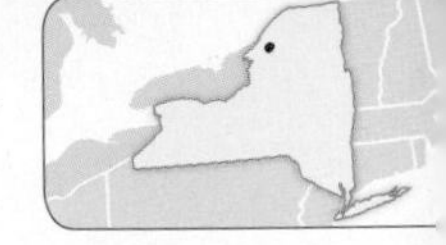

Primary Occurrence

ENVIRONMENT: Mines

WHAT TO LOOK FOR: Light-colored, fluorescent masses or tiny grains in rock from the Balmat-Edwards area

SIZE: Grains of turneaureite are smaller than ¼ inch while masses may be several inches

COLOR: Colorless to white or gray; fluorescent orange under shortwave UV

OCCURRENCE: Very rare

NOTES: Alongside the extremely rare donpeacorite (page 89) of the Balmat-Edwards area in northern New York, you'll typically find turneaureite. An arsenic-rich member of the apatite group (page 47), it only occurs in a handful of places around the world and in only one locality in the country, the ZCA No. 4 Mine, where it was procured from rock a half mile down into the earth. It doesn't appear as well-formed crystals, but instead occurs as small grains or masses, often clustered together and is typically colorless or white, making it very easily overlooked. But when placed under a shortwave ultraviolet light, specimens take on a completely new appearance as they glow vividly in shades of orange and yellow. Since turneaureite is found nowhere else in the state, its fluorescent habit is generally the only observation needed to identify it. Aside from donpeacorite, it is generally found in its host marble with amphiboles (page 39) and certain tourmalines (page 227).

WHERE TO LOOK: The ZCA No. 4 Mine in the Balmat-Edwards mining district of St. Lawrence County is the only locality in New York and one of only two on the continent. The mine is closed and largely off-limits, but material may still be available in the mine dumps if permission can be obtained to collect there.

Cluster of uralite crystals
Square, blocky shapes
Square cross section
Fine uralite crystal with wedge-shaped tip

Uralite

HARDNESS: 5-6 **STREAK:** White to gray

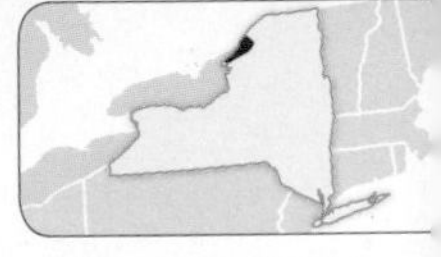

Primary Occurrence

ENVIRONMENT: Mines, hills, outcrops

WHAT TO LOOK FOR: Rare, elongated rectangular crystals grown in clusters; often pitted and rough in appearance

SIZE: Individual crystals are frequently an inch or longer, but clusters may be fist-sized or larger

COLOR: Dark green to gray or black

OCCURRENCE: Very rare

NOTES: Uralite was once considered a distinct mineral until it was better understood, and now we know that it is actually a particularly unique variety of amphibole (page 39). "Uralite" is the name given to pseudomorphs of an amphibole, usually actinolite (page 37) after a pyroxene (page 181), particularly diopside (page 85) or augite. A pseudomorph occurs when a mineral is chemically altered and changed into a completely different mineral on a molecular level while still retaining the outward crystal appearance of the original mineral. As such, uralite features pyroxene crystal forms, typically appearing as elongated rectangular prisms with square cross sections, and it is often intergrown in irregular clusters. The crystals are opaque and often have rough or pitted surfaces, all owing to their transformation, and most are colored in muted shades of dark green. This appearance is rather distinctive, but its hardness and typical occurrence with calcite (page 63) or feldspars (page 93) will also aid identification. Uralite is rare in New York, however, so finding it will be challenging.

WHERE TO LOOK: Only two significant localities exist in New York for this rare material: granite outcrops in the Alexandria Bay area and the metamorphic rocks of the Pierrepont area have produced the best specimens.

Complex vesuvianite crystal

Complex vesuvianite crystal

Vesuvianite

HARDNESS: 6.5 **STREAK:** White

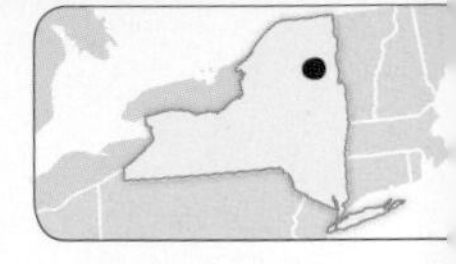

Primary Occurrence

ENVIRONMENT: Mines, outcrops, hills

WHAT TO LOOK FOR: Fairly hard, translucent, glassy dark crystals with a square cross section embedded in metamorphic rocks

SIZE: Crystals are generally no larger than a thumbnail

COLOR: Brown to red-brown, dark yellow to greenish

OCCURRENCE: Very rare

NOTES: Named for Mt. Vesuvius in Italy where it was first discovered, vesuvianite is a very complex silicate mineral that forms only in metamorphic rocks, particularly skarn and other metamorphosed limestones. Its hard, glassy, translucent crystals are prisms with an often somewhat blocky shape, with more-or-less square cross sections and low-angled points on each end, though in New York's vesuvianite-producing localities the crystals' corners are often slightly rounded, as if "melted." Crystals can be clustered together in parallel groups or intergrown with some crystals penetrating others, and frequently occur with scapolite (page 193) and diopside (page 85). But it is quite rare in the state, found in only a handful of sites, and with crystals rarely larger than a half-inch, they can be difficult to find. Identification, however, isn't quite as challenging, as vesuvianite's high hardness and characteristic occurrence within coarse-grained metamorphic rocks are distinctive. Poorly formed specimens may resemble various garnets (page 115), but garnets are much more common and are generally harder.

WHERE TO LOOK: Most specimens of collectible size have originated in Essex County, particularly the Minerva area around Olmstedville where outcrops and diggings have produced small but gemmy crystals with scapolite.

Well-formed warwickite crystal (⅛") in marble
Warwickite grains
(black)
Chondrodite
Crude warwickite crystals
(black) in marble

Warwickite

HARDNESS: 5.5-6 **STREAK:** Blue-black

Primary Occurrence

ENVIRONMENT: Mines, quarries, outcrops, hills

WHAT TO LOOK FOR: Tiny, black, dull prisms embedded in marble or other metamorphic rocks from the Warwick area

SIZE: Crystals are tiny, less than ¼ inch and usually smaller

COLOR: Gray to black, dark brown

OCCURRENCE: Very rare

NOTES: Warwickite is one of New York's type locality minerals, meaning that it was first discovered in the state, specifically in the town of Warwick for which it is named. A rare boron-bearing mineral that forms almost exclusively in marble, warwickite develops as tiny elongated prisms. Its crystals are rectangular and blocky in shape with subtle tips on each end that are often rounded and poorly defined. They are nearly always dark gray to black with a variable luster, but typically dull, though fine specimens may be nearly metallic in appearance. Their small size makes them easily overlooked and somewhat difficult to identify, but the white marble in which they occur contrasts well with their dark color and they frequently form alongside chondrodite (page 135) and spinel (page 215), which is a distinctive association. In fact, dark spinels may be a source of confusion, though they have a very different shape and are much harder than warwickite. Graphite (page 121) could also look similar, but is much softer, and magnetite (page 155) shares a similar color but is magnetic.

WHERE TO LOOK: The Amity-Warwick area quarries and outcrops of Orange County are the area of original discovery and the only place where specimens are still likely to be found, not only in New York, but in the entire country. Some areas are off-limits to collectors, but specimens are regularly available for sale.

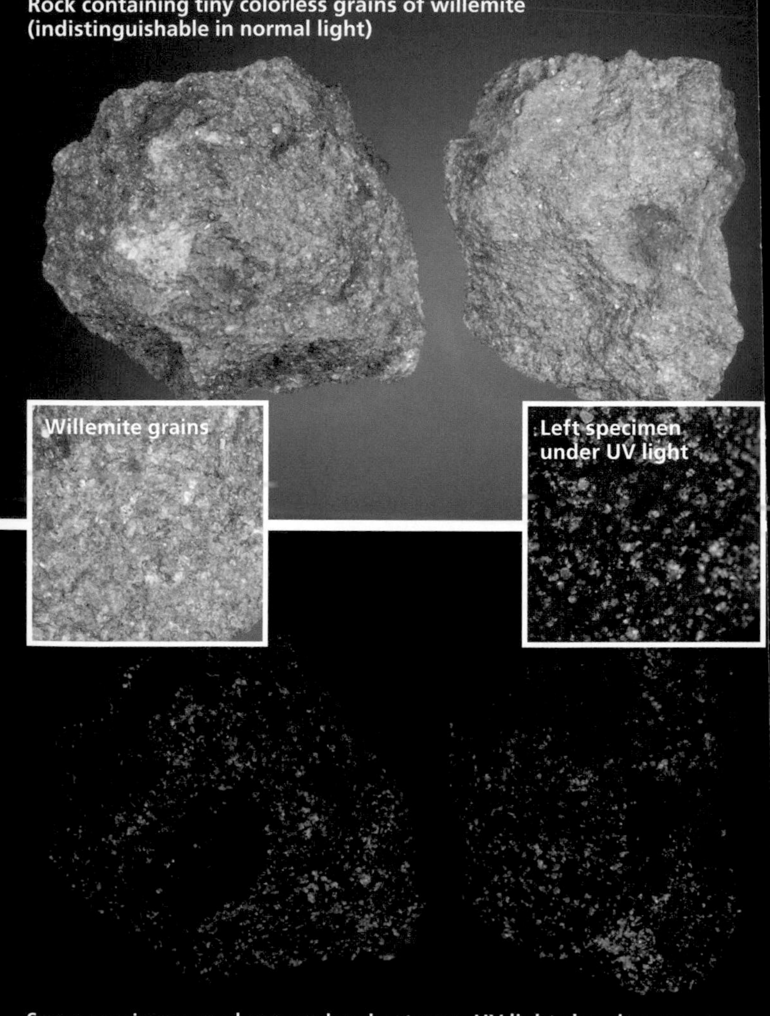

Same specimens as above under shortwave UV light showing green willemite fluorescence

Willemite

HARDNESS: 5.5 **STREAK:** White

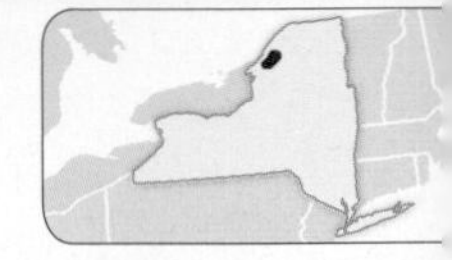

Primary Occurrence

ENVIRONMENT: Mines, outcrops

WHAT TO LOOK FOR: Tiny glassy grains that fluoresce green under UV light in zinc-bearing rocks

SIZE: Individual grains are tiny, no more than ⅛ inch

COLOR: Colorless to white or gray; fluorescent green under UV light

OCCURRENCE: Very rare

NOTES: In New York, willemite is a mineral that can go completely overlooked unless it's viewed under a specific kind of light. That's because it is highly fluorescent; fluorescent minerals absorb ultraviolet (UV) light radiation and re-emit it as visible light, often in striking colors. While other nearby states produce willemite's stubby, prismatic crystals, these are virtually unknown from New York, where it instead appears as tiny indistinct glassy grains, barely distinguishable from its host rock and nearby minerals. This is where a shortwave UV lamp will help, as the hidden willemite grains will glow brightly in their typical bright green color when exposed to UV light. Willemite forms as an alteration product of zinc-bearing minerals, particularly sphalerite (page 213), and though zinc ores are abundant in areas of New York, willemite isn't common anywhere. Because of its indistinct appearance under normal light, willemite is difficult to identify; under UV light, its vivid fluorescence is typically enough for a tentative identification, at least.

WHERE TO LOOK: Willemite is a rewarding mineral for collectors of fluorescent minerals, but in New York it was only recovered from the Balmat-Edwards zinc district of St. Lawrence County. Material is rare today from the long-closed mines, but with permission, some mine dumps may still be searchable.

Large, pure wollastonite crystal cluster from skarn deposit

Step-like cleavage

Fibrous luster

Crystal mass detail

Wollastonite-rich skarn (virtually all lustrous white grains are wollastonite)

Fibrous intergrown crystal masses

Wollastonite

Primary Occurrence

HARDNESS: 4.5-5 **STREAK:** White

ENVIRONMENT: Mines, quarries, outcrops, mountains, hills

WHAT TO LOOK FOR: Masses of fairly hard, brittle, fibrous and often fluorescent material in metamorphic rocks

SIZE: Specimens can range greatly, up to several feet

COLOR: White to gray, tan to brown, occasionally greenish

OCCURRENCE: Uncommon

NOTES: Wollastonite, a calcium silicate mined for use in ceramics and industrial applications, is a fairly abundant mineral worldwide and is widespread in New York. In New York it formed primarily within metamorphic rocks such as skarn. A light-colored, translucent mineral, wollastonite typically forms as thin, elongated crystals that grow in large masses; the thin crystals give those masses a distinctly fibrous appearance, sometimes in a radial arrangement. Specimens are often dotted with other metamorphic minerals such as garnets and diopside, and they are frequently intergrown with large amounts of calcite, with which it is easily confused. But wollastonite is far less common than calcite and is considerably harder, and though wollastonite has good blocky cleavage (breaks in square, blocky, step-like shapes), it should be easily distinguishable from calcite's perfect rhombohedral cleavage (breaks into perfect leaning-cube shapes). However, fibrous wollastonite tends to be brittle and will splinter, rather than cleave. When each mineral is tightly intergrown, however, hardness will be your best bet as well as the fact that calcite tends to be more translucent.

WHERE TO LOOK: The areas around Keene, Cascade Mountain, and Willsboro in Essex County, as well as northern Lewis County in Diana Township, have produced good specimens.

Tiny (approx. 1/16") chabazite crystals

Chabazite crystal (1/16")

Pearly heulandite crystal

"Coffin-shaped" crystal

Heulandite crystal cluster

Zeolite group

HARDNESS: 3.5-5.5 **STREAK:** Colorless to white

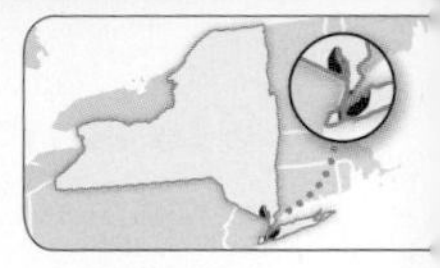

Primary Occurrence

ENVIRONMENT: Hills, lowlands, quarries, mines, rivers

WHAT TO LOOK FOR: Small, brittle, light-colored crystals within cavities in dark rocks

SIZE: Most zeolite minerals form crystals smaller than ¼ inch

COLOR: Colorless to white, pinkish, light brown

OCCURRENCE: Rare to very rare, depending on mineral

NOTES: The zeolites are a large family of minerals that form in cavities within rocks, particularly dark igneous rocks like basalt and diabase. They generally form when their host rocks weather and are affected by mineral-bearing groundwater. Only a few group members are present in New York. These include chabazite, heulandite, and stilbite; none are common, as their host rocks are also uncommon in the state. Zeolites are found in cracks and vesicles (gas bubbles) and crystals tend to remain small, their shapes differing from mineral to mineral. Chabazite, for example, forms as rhombohedrons (a shape like a leaning cube) resembling calcite (page 63), while heulandite develops elongated crystals widest at their center, appearing "coffin-shaped." Stilbite develops as thin, needle-like crystals, often arranged into radial "sprays." In most cases, though, zeolites are light-colored, lustrous, and fairly soft. Distinguishing them from other more common minerals generally takes additional research, though most casual collectors won't find New York's rare zeolites.

WHERE TO LOOK: Some of the best zeolite specimens from the state were found during construction beneath New York City; obviously, these are unobtainable today. Other specimens turn up infrequently in basalt of Rockland County.

Glossary

AGGREGATE: An accumulation or mass of crystals

ALKALINE: Describes substances containing alkali elements such as calcium, sodium, and potassium; having the opposite properties of acids

ALTER: Chemical changes within a rock or mineral due to the addition of mineral solutions

AMPHIBOLE: A large group of important rock-forming minerals commonly found in granite and similar rocks

ASSOCIATED: Minerals that often occur together due to similar chemical traits

BAND: An easily identified layer within a mineral

BED: A large, flat mass of rock, generally sedimentary

BOTRYOIDAL: Crusts of a mineral that formed in rounded masses, resembling a bunch of grapes

BRECCIA: A coarse-grained rock composed of broken angular rock fragments solidified together

CHALCEDONY: A massive, microcrystalline variety of quartz

CLEAVAGE: The property of a mineral to break along the planes of its crystal structure, which reflects its internal crystal shape; referred to in terms of shape or angles

COMPACT: Dense, tightly formed rocks or minerals

CONCENTRIC: Circular, ringed bands that share the same center, with larger rings encompassing smaller rings

CONCHOIDAL: A circular shape, often resembling a half-moon; generally referring to fracture shape

CRUST: The rigid outermost layer of the earth

CRYSTAL: A solid body with a repeating atomic structure formed when an element or chemical compound solidifies

CUBIC: A box-like structure with sides of an equal size

DEHYDRATE: To lose water contained within

DETRITUS: Debris, especially plant matter

DRUSE: A coating of small crystals on the surface of another rock or mineral

DULL: A mineral that is poorly reflective

EARTHY: Resembling soil; dull luster and rough texture

EFFERVESCE: When a mineral placed in an acid gives off bubbles caused by the mineral dissolving

ERUPTION: The ejection of volcanic materials (lava, ash, etc.) onto the earth's surface

FACE: A typically smooth surface of a crystal derived from a mineral's growth structure

FELDSPAR: An extremely common and diverse group of light-colored minerals that are most prevalent within rocks and make up the majority of the earth's crust

FIBROUS: Fine, rod-like crystals that resemble cloth fibers

FLUORESCENCE: The property of a mineral to give off visible light when exposed to ultraviolet light radiation

FRACTURE: The way a mineral breaks or cracks when struck, often referred to in terms of shape or angles

GEODE: A hollow rock or mineral formation, typically exhibiting a very round, ball-like external shape and interior walls lined with minerals, namely quartz and calcite

GLASSY: A mineral with a reflectivity similar to window glass, also known as "vitreous luster"

GNEISS: A rock that has been metamorphosed so that some of its minerals are aligned in parallel bands

GRANITIC: Pertaining to granite or granite-like rocks

GRANULAR: A texture or appearance of rocks or minerals that consist of grains or particles

HEXAGONAL: A six-sided structure

HOST: A rock or mineral on or in which other rocks and minerals occur

HYDROUS: Containing water

IGNEOUS ROCK: Rock resulting from the cooling and solidification of molten rock material such as magma or lava

IMPURITY: A foreign mineral within a host mineral that often changes properties of the host, particularly color

INCLUSION: A mineral that is encased or impressed into a host mineral

IRIDESCENCE: When a mineral exhibits a rainbow-like play of color, often only at certain angles

LAVA: Molten rock that has reached the earth's surface

LUSTER: The way in which a mineral reflects light off of its surface, described by its intensity

MAGMA: Molten rock that remains deep within the earth

MASSIVE: Mineral specimens found not as individual crystals but rather as solid, compact concentrations; rocks are often described as massive; in geology, "massive" is rarely used in reference to size

MATRIX: The rock in which a mineral forms

METAMORPHIC ROCK: Rock derived from the alteration of existing igneous or sedimentary rock through the forces of heat and pressure

METAMORPHOSED: A rock or mineral that has already undergone metamorphosis

MICA: A large group of minerals that occur as thin flakes arranged into layered aggregates resembling a book

MICROCRYSTALLINE: Crystal structure too small to see with the naked eye

MINERAL: A naturally occurring chemical compound or native element that solidifies with a definite internal crystal structure

NATIVE ELEMENT: An element found naturally uncombined with any other elements, e.g., copper

NODULE: A rounded mass consisting of a mineral, generally formed within a vesicle or other cavity

OCTAHEDRAL: A structure with eight faces, resembling two pyramids placed base-to-base

OPAQUE: Material that lets no light through

ORE: Rocks or minerals from which metals can be extracted

OXIDATION: The process of a metal or mineral combining with oxygen, which can produce new colors or minerals

PEARLY: A mineral with reflectivity resembling that of a pearl

PEGMATITE: The deepest portion of a granite formation, composed of large, interlocking crystals. Extremely slow cooling caused the minerals within the rock to crystallize to very large sizes; often contains rare minerals

PLACER: Deposit of sand containing dense, heavy mineral grains at the bottom of a river or a lake

PRISMATIC: Crystals with a length greater than their width

PSEUDOMORPH: When one mineral chemically takes the place of another but retains the outward appearance of the initial mineral

PYROXENE: A group of hard, dark, rock-building minerals that make up many dark-colored rocks like basalt or gabbro

RADIATING: Crystal aggregates growing outward from a central point, often resembling the shape of a paper fan

REPLACEMENT: See *pseudomorph*

RHOMBOHEDRON: A six-sided shape resembling a leaning or skewed cube

ROCK: A massive aggregate of mineral grains

ROCK-BUILDER: Refers to a mineral important in rock creation

SCHILLER: A mineral that exhibits internal reflections or "flashes" from within its structure when rotated in bright light, often showing an interplay of white, yellow, or blue

SCHIST: A rock that has been metamorphosed so that most of its minerals have been concentrated and arranged into parallel layers

SEDIMENT: Fine particles of rocks or minerals deposited by water or wind, e.g., sand

SEDIMENTARY ROCK: Rock derived from sediment being cemented together

SILICA: Silicon dioxide. Forms quartz when pure and crystallized, and contributes to thousands of minerals

SPECIFIC GRAVITY: The ratio of the density of a given solid or liquid to the density of water when the same amount of each is used, e.g., the specific gravity of copper is approximately 8.9, meaning that a sample of copper is about 8.9 times heavier than the same amount of water

SPECIMEN: A sample of a rock or mineral

STALACTITE: A cone-shaped mineral deposit grown downward from the roof of a cavity; sometimes described as icicle-shaped. Formations in this shape are said to be stalactitic

STRIATED: Parallel grooves in the surface of a mineral

TABULAR: A crystal structure in which one dimension is notably shorter than the others, resulting in flat, plate-like shapes

TARNISH: A thin coating on the surface of a metal, often differently colored than the metal itself (see *oxidation*)

TRANSLUCENT: A material that lets some light through

TRANSPARENT: A material that lets enough light through as to be able to see what lies on the other side

TWIN: Two or more crystals intergrown within or through each other

VEIN: A mineral, particularly a metal, that has filled a crack or similar opening in a host rock or mineral

VESICLE/VESICULAR: A cavity created in an igneous rock by a gas bubble trapped when the rock solidified; a rock containing vesicles is said to be vesicular

VOLCANO: An opening, or vent, in the earth's surface that allows volcanic material such as lava and ash to erupt

VUG: A small cavity within a rock or mineral that can become lined with different mineral crystals

WAXY: A mineral with a reflectivity resembling that of wax

ZEOLITE: A group of similar minerals with very complex chemical structures that include elements such as sodium, calcium, and aluminum combined with silica and water and typically form within cavities in basalt as it is affected by mineral-bearing alkaline groundwater

New York Rock Shops and Museums

ACE OF DIAMONDS ("Herkimer diamond" mine)
84 Herkimer Street
Route 28
Middleville, NY 13406
315-891-3855
www.herkimerdiamonds.com

AMERICAN MUSEUM OF NATURAL HISTORY
Harry Frank Guggenheim Hall of Minerals
Central Park West at 79th Street
New York, NY 10024
212-769-5100
www.amnh.org

ASTRO GALLERY OF GEMS (mineral shop)
417 5th Avenue
New York, NY 10016
212-889-9000
www.astrogallery.com

CRYSTAL GROVE DIAMOND MINE & CAMPGROUND
161 County Highway 114
St. Johnsville, NY 13452
1-800-579-3426

GORE MOUNTAIN MINERAL SHOP/GARNET MINE TOURS
Barton Mines Road
North River, NY 12856
www.garnetminetours.com

NATURAL STONE BRIDGE & CAVES
535 Stone Bridge Road
Pottersville, NY 12860
518-494-2283

NEW YORK STATE MUSEUM (exquisite NY mineral exhibit)
260 Madison Avenue
Albany, NY 12230
518-474-5877
www.nysm.nysed.gov

PAST AND PRESENT FOSSIL AND ROCK SHOP
3767 South Park Avenue
Blasdell, NY 14219
716-825-2361
www.pastpres.com

ROCK CITY PARK (geological site)
505 Route 16
Olean, NY 14760
1-866-404-7625
www.rockcitypark.com

ROCKO MINERALS & JEWELRY (by appointment only)
438 Southside Spur
Margaretville, NY 12455
845-586-3978

TWIN CRYSTAL ROCK SHOP
36 Broadway Street
Saranac Lake, NY 12983
518-891-1714

Bibliography and Recommended Reading

Books about New York minerals

Jensen, David E. *Minerals of New York State*. Rochester: Ward Press, 1978.

Chamberlain, Steven C. and Robinson, George W. *Collector's Guide to the Minerals of New York State, The*. Atglen: Schiffer Publishing, Ltd., 2013.

Van Diver, Bradford. *Roadside Geology of New York*. Missoula: Mountain Press Publishing Company, 1985.

Also see: Three-part special report on New York's mineralogy presented by *Rocks & Minerals* magazine written by several experts in New York's geology.
Part 1: November/December 2007. Vol. 82, No. 6
Part 2: May/June 2008. Vol. 83, No. 3
Part 3: May/June 2009. Vol. 84, No. 3
www.rocksandminerals.org

Bibliography and Recommended Reading *(continued)*

General Reading

Bates, Robert L., editor. *Dictionary of Geological Terms, 3rd Edition.* New York: Anchor Books, 1984.

Bonewitz, Ronald Louis. *Smithsonian Rock and Gem.* New York: DK Publishing, 2005.

Chesteman, Charles W. *The Audubon Society Field Guide to North American Rocks and Minerals.* New York: Knopf, 1979.

Johnsen, Ole. *Minerals of the World.* New Jersey: Princeton University Press, 2004.

Mottana, Annibale, et al. *Simon and Schuster's Guide to Rocks and Minerals.* New York: Simon and Schuster, 1978.

Pellant, Chris. *Rocks and Minerals.* New York: Dorling Kindersley Publishing, 2002.

Pough, Frederick H. *Rocks and Minerals.* Boston: Houghton Mifflin, 1988.

Robinson, George W. *Minerals.* New York: Simon & Schuster, 1994.

Index

About the Authors

Dan R. Lynch has a degree in graphic design with emphasis on photography from the University of Minnesota Duluth. But before his love of art and writing came a passion for rocks and minerals, developed during his lifetime growing up in his parents' rock shop in Two Harbors, Minnesota. Combining the two aspects of his life seemed a natural choice and he enjoys researching, writing about, and taking photographs of rocks and minerals. Working with his father, Bob Lynch, a respected veteran of Lake Superior's agate-collecting community, Dan spearheads their series of rock and mineral field guides—definitive guidebooks that help amateurs "decode" the complexities of geology and mineralogy. He also takes special care to ensure that his photographs complement the text and always represent each rock or mineral exactly as it appears in person. He currently lives in Madison, Wisconsin, with his wife, Julie, where he works as a writer and photographer.

Bob Lynch is a lapidary and jeweler living and working in Two Harbors, Minnesota. He has been cutting and polishing rocks and minerals since 1973, when he desired more variation in gemstones for his work with jewelry. When he moved from Douglas, Arizona, to Two Harbors in 1982, his eyes were opened to Lake Superior's entirely new world of minerals. In 1992, Bob and his wife Nancy, whom he taught the art of jewelry making, acquired Agate City Rock Shop, a family business founded by Nancy's grandfather, Art Rafn, in 1962. Since the shop's revitalization, Bob has made a name for himself as a highly acclaimed agate polisher and as an expert resource for curious collectors seeking advice. Now, the two jewelers keep Agate City Rocks and Gifts open year-round and are the leading source for Lake Superior agates, with more on display and for sale than any other shop in the country.

Notes

Your must-have guide to the rocks and minerals of New York

Full-color photos and the details you need to know for identifying and collecting

- **incredible, sharp, full-color photos:** the authors know rocks, and they took their own photographs to depict the detail needed for identification—no need to guess from line drawings
- **comprehensive entries:** photos and facts for 105 rocks and minerals
- **common rocks and rare finds:** fascinating information about everything from garnets and "Herkimer diamonds" to fossils and labradorite
- **easy-to-use format:** quickly find what you need to know and where to look

Also available: *Rocks & Minerals Playing Cards* features 52 common and sought after specimens in the United States. *Rock Hound's Logbook & Journal* is the perfect way to record when and where you unearth each of your finds.

ISBN 978-1-59193-524-7

$14.95

Adventure Publications
Adventure Publications
820 Cleveland Street. South
Cambridge, MN 55008
1-800-678-7006
www.adventurepublications.net